William A. Tilden

**Hints on the Teaching of Elementary Chemistry in Schools
and Science Classes**

William A. Tilden

Hints on the Teaching of Elementary Chemistry in Schools and Science Classes

ISBN/EAN: 9783337163983

Printed in Europe, USA, Canada, Australia, Japan

Cover: Foto ©Paul-Georg Meister /pixelio.de

More available books at **www.hansebooks.com**

HINTS
ON
THE TEACHING
OF
ELEMENTARY CHEMISTRY
IN
SCHOOLS AND SCIENCE CLASSES

BY

WILLIAM A. TILDEN, D.Sc., F.R.S.

PROFESSOR OF CHEMISTRY IN THE ROYAL COLLEGE OF SCIENCE, LONDON
EXAMINER IN CHEMISTRY TO THE DEPARTMENT OF SCIENCE AND ART

SECOND EDITION

LONGMANS, GREEN, AND CO.
LONDON, NEW YORK, AND BOMBAY
1896

PREFACE

THE issue of the new Syllabus of Inorganic and Organic Chemistry by the Department of Science and Art marks a very important advance in the history of the teaching of these subjects in this country. In this Syllabus the treatment of Chemistry from the theoretical side remains necessarily at the discretion of the teacher, though from the order in which the subdivisions of the matter for treatment are placed, it is obvious that, in the early stages especially, it is considered desirable to keep theory in a subordinate position, and to make use of it only when a sufficient foundation of fact has been duly acquired by the pupil. According to the experience of the Author, it is scarcely possible to use a purely inductive method in dealing with young students; but, though this may be admitted, the necessity of clearly distinguishing fact from hypothesis requires to be established far more definitely than at present in the minds of a large proportion of the teachers whose pupils present themselves for the examinations of the Department.

In the earliest stages, the learner's attention and energy are usually wholly used up in the process of exact observation. To see and accurately record the whole of a given phenomenon is enough for young boys and girls, and, until sufficient practice and experience in this direction have been gained, it is a better educational exercise than the attempt to solve problems which, to render them simple enough, require the neglect of part of the phenomena. For example, the investigations of the changes which occur when a piece of sheet copper is heated in the air requires for a complete answer observations (a) on the iridescent colours first visible, (b) the black coating which succeeds, (c) the red lining to the scales which may be separated on cooling, (d) the temperature at which these changes occur, (e) the alteration in weight of the mass, (f) changes in the surrounding air, &c. Observation of all these facts is possible even for the beginner; but when he is in possession of the whole, he would be a wonderful boy indeed who could, without assistance, infer that the products contain oxygen as well as copper, and that there are two oxides of copper of definite composition, to say nothing about the colours of thin films.

Chemistry is a science which, in its modern form, has sprung up almost entirely within the memory of living man, and its boundaries, already far reaching, are being rapidly extended year by year, and even month by month, by the labours of an army of workers. This condition of growth constitutes a feature of the science which should render it as a study attractive in no ordinary degree ; and the teacher who desires to make use of it as an effective educational agent must devote some part of his time to extending and consolidating his own knowledge. Not that the very latest discovery is necessarily more important than many of the facts which have been long familiar, and the teacher ought to be careful to exercise due discretion in communicating to a class of young pupils the most recent results of observation, lest their sense of proportion should be unequal to the strain of distinguishing the important from the comparatively unimportant. The discovery of argon in the atmosphere twelve months ago possesses the greatest interest for the advanced student ; but oxygen still ranks higher in importance, from all practical as well as theoretical points of view, than the newly discovered gas.

In the new Syllabus issued by the Department the most important changes will be noticed in the part which relates to the practical examinations. Scientific chemistry is based almost wholly upon observation and experiment, and the practical study of this science is one of the most useful means of developing the faculties of observation. But even

vi *Teaching of Elementary Chemistry*

in the teaching of analytical chemistry, which is especially dependent upon the practice of careful and exact observation, the 'crammer' has been at work to such an extent as largely to destroy the educational value of this important and interesting subject, and, instead of being made to record accurately the results of his own observations, the pupil is too often allowed to learn by rote tables or schemes of analysis, and is led to state, not what he sees, but what the book tells him he ought to see. A form of examination question used by the Author which requires the candidate to describe some operation which he has himself conducted has often received answers so remarkable as to show that the faculty of observation is in some cases almost extinguished by the habit of blind reliance upon a printed text.

The following chapters contain the substance of a short course of Lectures given in July last to the class of Teachers assembled for instruction at the Royal College of Science ; and in the belief that the hints contained in them would be useful to others engaged in preparing for the May examinations, as well as to Teachers of Elementary Classes generally, they are now offered in a slightly amplified form, and with a little more detail than is possible in presenting the same subjects in the lecture-room.

ROYAL COLLEGE OF SCIENCE, LONDON :
September 1895.

CONTENTS

CHAPTER		PAGE
I.	OBSERVATION—QUALITATIVE	1
II.	OBSERVATION—QUANTITATIVE	12
III.	OBSERVATION—QUANTITATIVE	21
IV.	CHEMICAL EQUIVALENTS — THE LAW OF VOLUMES	31
	NOTE TO CHAPTER IV.	44
	CHRONOLOGICAL TABLE, FOUNDERS OF CHEMISTRY	42
V.	MOLECULAR WEIGHTS AND FORMULÆ	46
VI.	ATOMIC WEIGHTS—CLASSIFICATION	55
	APPENDIX	70

HINTS ON THE TEACHING

OF

ELEMENTARY CHEMISTRY

CHAPTER I

OBSERVATION—QUALITATIVE

THE history of chemistry affords numerous instances of fallacies originating as a consequence of imperfect observation. In some cases such fallacies have been received as articles of belief almost undisputed for centuries. The doctrine of the transmutability of metals, and especially of the base metals, into silver and gold is an example of this kind. It is a fact that almost any sample of lead, submitted to the process of cupellation, yields a little bead of silver which often contains gold. It was not an unnatural hypothesis that this silver somehow originated in a change brought about in the lead by the action of heat; and since in different operations the yield of precious metal was known to vary, it seemed likely that by suitably changing the conditions complete transmutation might be brought about.

The doctrine of Phlogiston, again, seemed to supply all that was wanted to explain the process of combustion and to bring under a common theory the phenomena of burning, rusting of metals, decay of wood, and fermentation of saccharine fluids. When a piece of wood or charcoal is burned, the theory taught that the escaping *phlogiston* was the cause of the appearance of fire, the *calx* remaining in the ashes upon the hearth. Metallic iron exposed to the air loses its phlogiston, and becomes transformed into a calx more copious than that of charcoal. The latter, then, is rich in phlogiston, and ought to be able to impart some of it to the calx of iron, and so restore it to the metallic or reguline condition, and this it does when iron ore or rust is heated with charcoal. All this is true as far as it goes, but the defect in the theory was that it did not take cognisance of the fact that when metals are calcined the resulting calx is heavier than the metal.

Another most instructive example of bad observation, and consequent erroneous theory, is exhibited in the history of the element we now call chlorine. This gas was discovered by Scheele in 1774, and called by him *dephlogisticated muriatic acid*. In this he was quite right, for such a designation means that the new gas consisted of muriatic acid deprived of its inflammable principle. Berthollet, however, concluded from his experiments that this body was composed of muriatic acid and oxygen, and for a long time it was thought that the hydrogen, which was undoubtedly procurable from muriatic acid by the action of metals upon the gas, was due to the accidental presence of water.

Sir Humphry Davy established the elemental character of the substance, and gave it the name *chlorine*, which refers solely to the yellowish-green colour of the gas, and not to any hypothesis as to its nature. But Davy succeeded only very gradually in getting rid of errors arising from the presence of moisture in the imperfectly dried gas. The principle adopted in his researches was the very simple one which consists in attacking the point in dispute directly. The question was whether 'oxymuriatic acid' contained oxygen, and this he answered by a series of experiments in which he proved that metals, charcoal, phosphorus, and other bodies heated in the gas give compounds recognisable as muriates, but never as oxides ; and he pointed out that the presence of oxygen or of hydrogen, or phlogiston, ' or of other principles, should not be assumed where they cannot be detected.'[1]

Among the chemists of the past there is none whose writings are more instructive, or whose fame

[1] For a complete account of Davy's work on Chlorine, see 'Alembic Club Reprints,' No. 9; W. F. Clay, Edinburgh.

is likely to be more enduring, than the industrious Scheele. And the reason of this is that his work was almost wholly experimental. As he himself says, 'Conjectures can determine nothing with certainty; at least they can only bring small satisfaction to a chemical philosopher who must have his proofs in his hands.'[1] Observation and experiment constitute the first and the most important business of all students of chemistry; but, that they may be fruitful, observation and experiment alike must be carried on not listlessly, carelessly, with 'lack-lustre eye,' but with concentrated attention and senses awake and active.

With all the accumulated experiences of the past, the art of observation is, however, still imperfectly learned, and the chemical journals teem with descriptions of phenomena or of substances which are either incomplete or inaccurate, or even wholly erroneous. Writers of text-books are also far from blameless in this matter, and the inability on the part of students to describe correctly even very simple things is often as much due to the perpetuation of false statements in the books as to want of attention or of skill on the part of the student. One of the most flagrant cases is the description of the usual process for the production of sulphur dioxide. A well-known text-book

[1] See Scheele's 'Chemical Treatise on Air and Fire,' of which the most interesting part, relating to the discovery of oxygen, is translated into No. 8 of the 'Alembic Club Reprints,' already referred to.

contains the following statement: 'This gas is prepared by the action of hot concentrated sulphuric acid upon copper turnings.

$$Cu + 2H_2SO_4 = CuSO_4 + 2H_2O + SO_2.'$$

That is the whole of the explanation given in this case. An observer of the process, however, cannot fail to see that, so soon as the action commences and the gas begins to escape, the copper becomes *black*; but the equation gives no explanation of this fact, and students are actually led to believe that the residue ought to be blue. According to the author's experience, candidates at examinations in which a question occurs as to the action of sulphuric acid upon copper do frequently make the statement that the residue is *blue*. Nor are students alone liable to fall into this error. A teacher once seriously proposed in the author's presence to assist the memory by writing chemical equations in chalks coloured so as to recall the appearance of the several materials concerned. Selecting the equation referred to by way of illustration, he wrote the Cu in red, the H_2SO_4 in white, and the $CuSO_4$ in blue chalk, ignorant or forgetful of the fact that $CuSO_4$ is a white substance.

The difficulty of making exact observations of comparatively simple facts is daily illustrated by the reports of the law courts, where different witnesses, actuated, we must believe, in many cases by no dishonest motive, often give wholly contradictory testimony as to occurrences of the most ordinary kind. Most people are also usually incapable of discriminating objects unless they differ enormously in size or colour. This is illustrated by the common names which have been applied for centuries to many familiar plants. There is, for example, the greater celandine and the lesser celandine, which have little resemblance to each other beyond the colour of the flowers. The plants called water-lilies, again, have no botanical connection with true lilies, neither do the flowers resemble those of the lily. For a long time there was, and there probably still exists, confusion in the minds of the majority of people as to the two forms of electric light and the 'incandescent' gas-burners now so common. Any kind of bright light seems to be sufficient to deceive the untrained eye.

In order to cultivate the powers of observation, various branches of natural science have been brought into use in schools, but none seem to present so many advantages as are offered by chemistry when rightly taught. As a science based entirely upon the results of observation and experiment, it is only by making experiment a principal feature of the system of instruction that these advantages can be secured. The observations and experiments must also, as far as possible, be the work of the pupil and not of the teacher, and therefore exercises undertaken should be in the first instance of the simplest possible character, and graduated so as to lead on to more difficult operations, which should only be undertaken after some time and after demonstration by the teacher. It is a mistake to suppose that the great theories of chemistry can be established by experiments conducted wholly by beginners, but with due preliminary instruction the more advanced student may get a long way in this direction.

The object of the following series of exercises is merely to suggest to teachers the kind of practical work which may be advantageously done by the pupil, and to show in a general way how he may proceed from very simple operations to work of a more complex character.

The earliest examples provide material for observation without requiring more than care and attention. The teacher will have no difficulty in devising an extended and almost infinite variety of such simple exercises. It will be necessary to insist upon a written account of each experiment made by the student, with a statement in his own words of what he sees, without at first requiring any theoretical explanation or discussion. It will often be advantageous to supply materials with which the student is not already familiar. When his observations have been successfully recorded, the teacher may, if he thinks proper, tell him the composition of the substance and explain the nature of the changes which he has noticed,

EXPERIMENTS TO BE DONE BY EACH PUPIL

I. *Observation of the action of heat* supplied by the gradual application of a Bunsen flame to a dry test-tube containing a little of the substance.

Any of the following or other substances may be tried: Mercuric oxide, red lead, lead nitrate, potassium chlorate, potassium nitrate, ammonium nitrate, ammonium chloride, mercuric iodide.

Gases evolved may be tested for as follows:

(1) Notice colour, appearance of fumes, odour.

(2) Apply lighted wax taper first at the mouth of the tube, then pushed inside.

(3) Hold in the gas strips of red and blue litmus paper, moistened with water.

(4) Hold within the tube a drop of clear lime water at the end of a narrow glass tube, then gently suck for a moment at the open end of the tube so as to draw up the drop.

This series of tests is not intended to enable the pupil to identify the gas, for that is a matter of comparatively small importance at this stage. After observations have been correctly recorded, the teacher may suggest to the pupil other experiments which appear to him desirable, such as the collection of the gas in bulk or further examination of the residue.

Some of the substances mentioned in the above list will afford opportunities of testing the genuineness and completeness of the student's work. For example, he may have read or learned that mercuric oxide gives off oxygen and mercury when heated, but unless he has tried the experiment or seen it tried, he could not guess that the powder would become black while hot, and that at a high temperature it would give a yellowish-coloured gas and a small yellow sublimate upon the sides of the tube, owing to the trace of nitrate which the commercial red oxide invariably contains.

II. *Crystallisation of salts from water*, and recognition of crystalline form, or at least differentiation of one sort of crystal from another.

For example, make strong solutions of potassium nitrate and ammonium chloride in hot water, and pour into separate watch-glasses. On cooling the long prisms of the nitre are easily distinguished from the fern-leaf forms of the sal ammoniac. Similarly alum, lead nitrate, and barium nitrate will yield regular octahedrons and dodecahedrons, and intermediate forms distinct enough to be readily visible and easily sketched. Other salts which crystallise from hot water are potassium chlorate, sodium nitrate, magnesium and zinc sulphates, copper sulphate, chrome alum, potassium dichromate, &c.

III. *Precipitates to be distinguished* by differences of density or consistence. The following are white precipitates differing in character: solution of alum mixed with ammonia gives a gelatinous precipitate; calcium chloride with carbonate of ammonia a flocculent precipitate which becomes sandy on heating or after standing; calcium chloride with dilute sulphuric acid a mass of minute crystalline needles; barium chloride with dilute sulphuric acid a fine powder some of which may remain suspended a long time; silver nitrate with a chloride a white precipitate which curdles and on exposure to daylight assumes a purple tint, &c.

In like manner arsenious sulphide, lead chromate, zinc chromate, stannic sulphide, silver iodide are examples of precipitates which have a yellow colour, but differ in tint and density.

IV. *The action of strong sulphuric acid upon substances* resulting in the evolution of gas.

The following examples will serve: common salt, potassium nitrate, red lead, potassium iodide, sodium sulphite, sodium formate.

A few grams of the substance may be placed in a test-tube and strong sulphuric acid added in quantity sufficient to make a fluid paste. If heat is applied it will be well to caution young students as to the corrosive nature of the liquid, and direct them to turn the open mouth of the tube away from their own faces and from the persons of their neighbours. The evolved gas may be tested as already described under Section I.

In addition to such exercises as the foregoing, specimens of well-crystallised compounds such as potassium iodide, alum, sugar, zinc sulphate, or of minerals such as galena, fluorspar, pyrites, calcite, or specular iron may be given to be drawn and described. And when qualitative analysis is begun, the pupil should be taught *in every case to write down, before applying tests, an account of the appearance and more obvious characters of the substance analysed.*

CHAPTER II

OBSERVATION—QUANTITATIVE

QUANTITATIVE experiments may be gravimetric or volumetric; that is, they may take the form of measurements of weight or of volume. Contrary to general belief, quantitative work may be done with very simple and inexpensive appliances, and in the hands of even young students results may be obtained which closely accord with theory. It is, however, advisable in the first instance to set exercises to be done independently of theory until a series of experiments has given concordant results, showing that the pupil has acquired sufficient skill to be allowed to test for himself some general law which he has already learnt.

The balance shown in the figure may be obtained of several instrument makers for about 27s. 6d., and a box of weights ranging from 100 grams to 1 milligram costs 6s.

The simplest quantitative experiments are those in which a gain or a loss of weight of a single object, such as a crucible, has to be recorded. The following are examples of the kind of experi-

Observation—Quantitative

ment referred to. They can all be done by beginners with very slight supervision and with satisfactory results. It is not necessary to weigh to a smaller subdivision than ·01 gram.

I. *Formation of metallic oxides and sulphides.*— Magnesium may easily be converted into the oxide by ignition in air, but to avoid loss it must be heated in a covered crucible. A clean 'half-ounce'

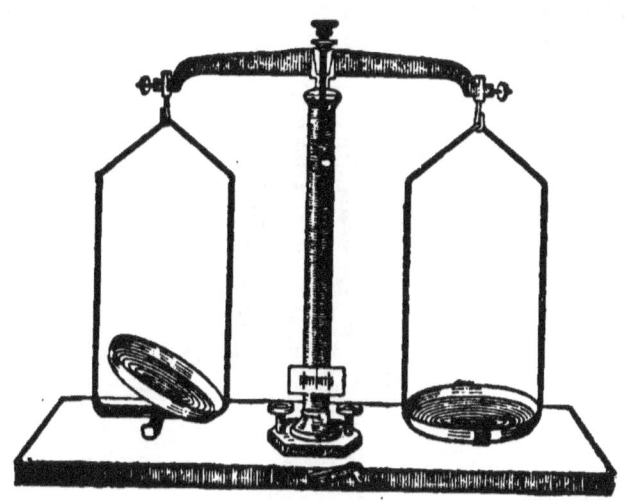

FIG. 1

covered porcelain crucible should be placed on the left-hand pan of the balance, and counterpoised by putting small shot into the lid of a pill-box placed in the opposite pan. About 15 centimetres of magnesium ribbon rubbed clean with sand-paper is then coiled up in the crucible, and its weight noted. It should weigh ·3 to ·4 gram. The crucible is then placed on a wire and tobacco pipe triangle supported on a tripod or retort stand ring over a Bunsen burner. The flame is then gradually applied till the bottom of the crucible is red-hot. The lid may be lifted slightly from time to time, and in about a quarter of an hour the glowing of the metal due to combustion will have ceased and a grey mass will remain. The lid should then be removed and a strong heat applied a little longer till the mass becomes white. 1 gram of magnesium gives 1·66 gram of the oxide. In a

FIG. 2

test experiment conducted as above, ·39 gram gave ·65 gram of oxide. This is in the proportion of 1 to 1·66.

Copper may be readily converted into the oxide; but as a simple roasting process would occupy too much time, it is best to convert the metal first into nitrate, which is decomposed by subsequent heating. Counterpoise a crucible as before, weigh out about half a gram of thin sheet or wire, add to it five or six drops of ordinary strong nitric acid, cover

Observation—Quantitative

the crucible and set it in a warm place till a dry green mass remains. Then heat over the Bunsen flame gradually to redness. The resulting oxide should of course be black and show no signs of unchanged metal. Should any unchanged metal be suspected, a few more drops of nitric acid should be added and the process repeated. Iron and tin may be dealt with in the same way.

    1 gram of copper gives 1·25 gram oxide
    1 gram of iron   „   1·43 „   „
    1 gram of tin   „   1·27 „   „

Copper may also be converted into sulphide, by heating it with sulphur. A counterpoised covered crucible containing about 1 gram of flowers of sulphur is placed over the Bunsen flame and heated strongly till the vapour of sulphur is seen escaping round the lid. A weighed piece of copper (about ·5 gram) is then dropped into the crucible, the lid being lifted only slightly and immediately replaced. The heating is continued till the excess of sulphur is completely volatilised. The crucible is then allowed to cool, and when cold is weighed. A small quantity of sulphur should be added to the contents of the crucible, which should be rapidly heated again to redness, cooled and reweighed. The second weight should be the same as the first.

For the success of this experiment the air must be excluded as much as possible from the crucible.

The lid must therefore not be removed while the sulphide contained in the crucible is still hot. A good way of preventing access of air is to finish the heating in an atmosphere of coal gas supplied through a tobacco pipe, attached to a rubber tube, and having the bowl inverted over the crucible. The escaping gas burns with a luminous flame round the top of the crucible. 1 gram of copper should theoretically give 1·25 gram of sulphide.

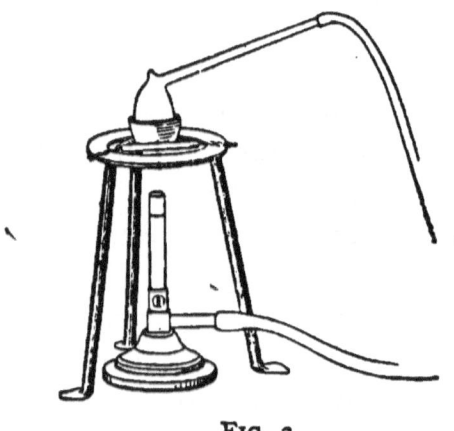

FIG. 3

An experiment conducted as described gave 1·26 gram.

II. *Decomposition of compounds by heat.*—Proceeding in a similar manner, quantitative estimations may be made of the products of various chemical decompositions produced by heat in which a fixed residue remains. For example, potassium chlorate, heated, leaves potassium chloride; magnesium carbonate and zinc carbonate leave a residue of oxide; crystallised sulphates of copper and other metals lose their water and leave anhydrous sulphates. Few remarks are necessary in connection with such experiments as these. Salts which are liable to crackle or decrepitate must of course be heated in a closely covered crucible. Copper sulphate, magnesium sulphate, zinc sulphate, and gypsum require to be heated only very gently by means of a rose burner placed some three inches below the bottom of the crucible. Commercial carbonates of zinc and magnesium have a nearly constant composition corresponding respectively to the formulæ $ZnCO_3. Zn(HO)_2. H_2O$ and $3MgCO_3. Mg(HO)_2. 4H_2O$.

One gram of crystallised copper sulphate should yield ·64 gram of anhydrous sulphate. An experiment gave 1·39 gram, from 2·16 grams, which corresponds with theory. The residue was slightly grey.

By such exercises as the foregoing, varied and repeated according to the discretion of the teacher, the student may be shown that he can for himself verify the fundamental law of chemistry—namely, the *Law of Definite Proportions*, which asserts that every chemical compound has a fixed and definite composition. The experimental method may after these exercises be modified so as to show that the same results may be obtained though the materials pass through a different series of processes.

III. *Treatment of precipitates.*—As an example

a piece of clean iron wire not more than ·5 gram in weight may be dissolved in a half-pint beaker by a few cubic centimetres of a mixture of hydrochloric with a little nitric acid. The solution is then diluted with water and excess of solution of ammonia added, the precipitate filtered off, the filter drained, dried, and burnt to ashes in a counterpoised crucible. The filter ash may be neglected. By this process also it may be shown that 1 gram of iron gives 1·43 gram of the red oxide.

In order to prove to the pupil that it is usually necessary to wash precipitates, the same experiment may be performed with the substitution of caustic soda for ammonia. After weighing the oxide of iron, add some boiling water to the contents of the crucible, let the oxide settle, and pour off the water. After repeating this once or twice, dry the whole and weigh again. The weight will be slightly reduced, and nearer to the theoretical amount. The washings will turn red litmus paper blue.

IV. *Replacement of metals by one another.*—A clean piece of zinc foil immersed in a somewhat dilute (2 to 5 per cent.) solution of copper sulphate in water becomes coated immediately with a black or brown furry coating of metallic copper, which continues to be thrown down till all the zinc has dissolved. A few small bubbles of hydrogen sometimes escape, but these may be neglected, as they represent only an extremely minute quantity of zinc. In order to show the ratio between the

Observation—Quantitative

weight of zinc dissolved and copper deposited, the clean zinc must be weighed and placed in a small beaker containing the solution of copper, which must be used in excess—that is, it must remain blue after the zinc is dissolved. The action is over when on feeling about with a glass rod no particles of the zinc are encountered. It is advisable to warm the solution toward the end. The reddish precipitate of copper is allowed to settle, the blue solution containing zinc and copper sulphates poured off, and the precipitate brushed by means of a camel-hair brush into a counterpoised crucible. The washing is completed in the crucible, by adding some boiling water, which, after settlement of the copper, is poured off as completely as possible. As the pulverulent metal is liable to oxidise if heated in contact with the air, it should be dried rapidly, and this may be accomplished by adding to it a few cubic centimetres of strong alcohol, pouring this away, and then placing the crucible in a steam oven for a few minutes. It is weighed when cold. Theoretically 1 gram of zinc precipitates ·97 gram of copper. By experiment conducted in the manner described, ·57 gram of zinc gave ·55 gram copper, which is in the proportion of 1 to ·965.

By a similar method of procedure:
- 1 gram magnesium precipitates 2·64 grams of copper.
- 1 gram copper precipitates 3·40 grams of silver.

1 gram magnesium precipitates 9 grams of silver
1 gram zinc „ 3·32 „ „ „

In the case of replacing silver by copper magnesium or zinc, an excess of solution of silver nitrate may be used.

From results obtained in this way the *Law of Reciprocal Proportions or Equivalents* may be tested by the pupil for himself. He finds, for example, by successive experiments, and calculating from the results, that 1 part of magnesium, 2·71 parts of zinc, and 2·62 parts of copper are respectively required to precipitate 9 parts of silver. From this he may infer that 2·71 parts of zinc and 2·62 parts of copper are equivalent to each other, or are capable of doing an equal amount of chemical work. This he can verify by recalculating the result of his experiment in which copper is precipitated by zinc, for

$$1 : ·97 :: 2·71 : 2·62.$$

Similarly he may infer that 1 part of magnesium is equivalent to 2·71 parts of zinc, though this cannot be similarly verified by direct precipitation. In a later chapter an experiment will be described by which the weight of hydrogen corresponding to these quantities of the metals can be experimentally determined; but as this requires more careful manipulation, and the observation of weight to the third place of decimals, it is not within the capacity of pupils at this stage.

CHAPTER III

OBSERVATION—QUANTITATIVE

Measurement of gases—In the preceding chapter the experiments described involve determinations of gain or loss of weight. The pupil may now proceed to make estimations of the volume of gas obtained in various processes, and it is surprising what close approximations to theoretical results are possible by the use of the simplest apparatus and by methods which appear to be among the least exact. Estimations of the volume of hydrogen evolved from acids by various metals afford the best exercises to begin upon, and at first all calculations may be neglected, inasmuch as the solubility of hydrogen in water and the difference between ordinary atmospheric conditions and normal temperature and pressure have comparatively little effect on the volume of gas collected. The apparatus shown in the figure may be recommended for its simplicity. It consists of a piece of tubing about 20 cm. long, 6–8 mm. internal diameter, drawn out at the top and bent into a curve, upon which is fitted, by means of a short piece of rubber tubing, a second piece of quill glass tubing, turned up at the end so as to convey the gas into a jar. The metal, zinc foil, or magnesium ribbon is thrust into the wider tube, and, to prevent portions of it from falling out of the open end, a little plug of copper gauze is stuffed into the mouth.

To collect the gas a common stoneware pneumatic trough, with beehive shelf and cylindrical gas

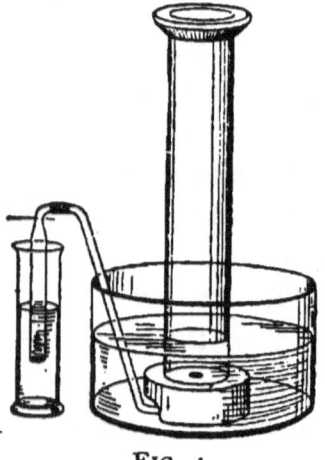

FIG. 4

jar filled with water, may be used. To make the experiment, the upturned end of the delivery tube is put under the mouth of the jar, and a deep beaker or cylinder containing dilute sulphuric acid is brought under the tube containing the metal, and the latter is slowly lowered into it. Evolution of gas commences immediately, and the gas passes into the jar. When the whole of the metal is dissolved, the tube may be removed and the level of

Observation—Quantitative

the water in the jar is marked off by means of a strip of paper label, which is affixed to the outside of the jar, and the edge of which marks the surface of the water or of the gas. The jar may then be lifted out of the trough, water poured into it till the surface is level with the edge of the paper strip, and this water, which now occupies the space taken up previously by the gas, is measured by pouring it out into a graduated cylinder.

No attempt should be made to expel the whole of the gas from the tube into the jar, for this residual gas occupies the space previously filled by air, which at the beginning of the experiment was driven into the jar. When magnesium is used it should be lowered into the acid very cautiously or some bubbles of gas may be lost.

A series of experiments should in the first instance be made as described, and the resulting volume of gas recorded in the notebook; at the same time the temperature of the water and the height of the barometer may be noted, the corrections being reserved by beginners, and applied later, when they have learned the effect of changes of temperature and pressure upon gases, and can make the necessary calculations.

One gram of zinc gives theoretically 343 c.c. of hydrogen at normal temperature and pressure. An experiment in which ·86 of zinc foil was used gave 315 c.c. at 20°; 1 gram would, therefore, have given 366 c.c., or 341 c.c. at normal temperature.

One gram of magnesium gives, theoretically, 930 c.c. of hydrogen at normal temperature and pressure.

By experiment, ·38 gram of magnesium gave 385 c.c. of hydrogen at 20°: so that 1 gram would have given 1013 c.c., or 943 c.c. at normal temperature.

Here it may be remarked that the correction required for pressure is usually much less than the correction for temperature, and may very commonly be neglected altogether for purposes such as we have in view in these experiments.

It may be observed that the measurement of the gas evolved from acid by the metals may be made use of for the purpose of estimating the quantities of aluminium, zinc, and magnesium, which are equivalent to one another, by calculating from the result obtained the amount of metal required to yield the same amount, say 1000 c.c., of hydrogen.

Another good form of experiment involving the measurement of gas is the estimation of oxygen in atmospheric air by means of phosphorus. A graduated measuring cylinder may be used for this purpose, or simply a plain gas cylinder holding about 250 c.c. A scratch is made about 5 cm. from the mouth, and the capacity of the jar to this mark is ascertained by filling with water and measuring the water. The jar is then inverted in a water-trough, and the air admitted till the surface of the water inside stands at the mark. The air in the jar then occupies the bulk of the measured quantity of water.

A piece of clean phosphorus, the size of a pea, is now stuck at the end of a piece of copper wire, and thrust up into the jar till it is near the top. The whole is allowed to stand 6–8 hours, the phosphorus withdrawn, and the new level of the water marked off with a strip of gummed paper. The volume of the residual nitrogen can then be measured by means of water as before.

Measurement of liquids. Neutralisation of acids and alkalis.—The metric system of weights and measures is a subject included in the syllabus of the Science and Art Department, and every teacher who professes to meet the requirements of the Department will find it necessary to give occasional demonstrations of the relations between the several denominations of weight and measure. This is one of those subjects which cannot be acquired offhand, but demands repeated consideration and use to make it familiar. Whether it has already been learned or not by the pupils, it will be advisable at this stage to begin operations by allowing them to handle some of the common measuring vessels of different forms, and to note the relation of burettes to graduated cylinders, flasks, &c. Volumetric exercises in great variety may then be practised.

For the purposes now contemplated we may assume, without appreciable error, that fresh oil of vitriol, which has not been long exposed to the air, may be regarded as consisting of real sulphuric acid, H_2SO_4. This assumption will be convenient to the teacher, though it is unnecessary and undesirable to communicate at first any notions of this kind to the pupil. A standard solution of dilute sulphuric acid may be prepared by weighing out 49 grams of oil of vitriol in a small beaker, and pouring this into water, sufficient being used, together with the rinsings of the beaker, to make up 1000 c.c.

The next step is to make choice of indicators, and it will be found that litmus or cochineal is the most useful. The burette to be used may be divided into ·2 or ·1 c.c. With the aid of the solution of sulphuric acid, a number of experiments may be quickly made on the neutralising power of various common alkaline solutions, such as solution of potash, soda, or ammonia, or of solid alkaline substances such as sodium carbonate. The only special precaution to be taken in dealing with carbonates is that when litmus is used the liquid requires to be boiled in order to effect complete decomposition and expulsion of the carbonic acid. Details of the mode of operating are given in so many manuals that it is unnecessary to describe the common process of alkalimetry or acidimetry in this place.

Suppose now it is found that 50 c.c. of the sulphuric acid are required to neutralise

> a c.c. of solution of potash
> b c.c. of solution of soda
> or c grams of sodium carbonate.

We can apply either of these facts to ascertain the neutralising power of other acids. Thus a c.c. of solution of potash will be found to neutralise

> m c.c. of hydrochloric acid
> n c.c. of nitric acid
> or p grams of oxalic acid.

Then $a, b,$ or c will neutralise not only 50 c.c. of sulphuric acid, but $m, n,$ or p of the other acids, and so these quantities, $a, b, c, m, n, p,$ are respectively equivalent to one another. So important is it to secure due and exact attention to fact before permitting the introduction of theory, that in this part of the work again the author would impress upon teachers once more the advisability of requiring young pupils to practise the observations of neutralising power before they are allowed to use formulæ. When the mere manipulation has been mastered, the explanation may be given of the use of the specific quantity of sulphuric acid taken for the preparation of the standard solution, and the reactions may be expressed by equations. It will be well, however, to confirm the result expressed in the equation in a few cases by direct experiment. Thus the amount of HCl in a sample of hydrochloric acid having been determined volumetrically, as explained above, 50 c.c. of the liquid may be exactly neutralised by soda, and the solution evaporated to dryness in a small counterpoised porcelain dish heated by a water bath. The salt when dry may be weighed. Another more difficult process would be to place a weighed quantity of salt in a small flask connected with a bulb absorption apparatus containing distilled water. On adding excess of strong sulphuric acid to the salt and heating, the whole of the hydrochloric acid gas may be driven into the water, and the resulting solution tested volumetrically by alkaline solution of known strength.

For exercise in manipulation and observation without the interference of theory there is no better example than the application of the soap test to the estimation of 'hardness' in water. It is not necessary to describe this process here, as an account of it is to be found in nearly all books on analysis.

The following statement occurred in a paper sent up at the recent May examination (1895) 'Common salt is called sometimes sodium chloride, and this shows that it must contain chlorine by the name.' The childish character of such an answer may be thought to be exceptional; but though it is true that such a crude form of statement is not

common, the fact remains that names and formulæ
are often used as expressions of fact when they represent the thing that is to be proved. An illustration of this is almost certain to occur whenever
the examinees are requested to illustrate the law
of multiple proportions by referring to, say, the
oxides of nitrogen. The formulæ, having been
committed to memory, are at once produced, the
candidates being apparently unconscious that any
reference to them is equivalent to begging the
whole question. If, however, the problem is stated
in another form without supplying the names of
the elements concerned, students seem to be incapable of attacking the question, and it remains
usually unanswered. Suppose the question to
stand as follows: 'Two elements A and B combine
in the following proportions:

	A	B
I	96·28	3·72
II	92·83	7·17
III	89·62	10·38
IV	86·62	13·38

Show that these combinations comply with the
law of multiple proportions.'

All that has to be done is to make one of these,
say A, a fixed quantity, and to calculate the several
proportions of the second element which are com-
bined with this quantity, and then see whether
they stand in the simple relation to one another
required by the law. If in the above example we
calculate the amount of B which is in each com-

pound united with 100 parts of A, the figures stand as follows:

			A		B
I	.	.	100	:	3·86
II	.	.	100	:	7·72 or 3·86 × 2
III	.	.	100	:	11·58 or 3·86 × 3
IV	.	.	100	:	15·45 or 3·86 × 4

Q.E.D.

CHAPTER IV

CHEMICAL EQUIVALENTS—THE LAW OF VOLUMES

IN the last chapter experimental methods were described by which the volumes of hydrogen expelled from acids by the action of several different metals could be estimated, and hence the weight of the metals equivalent to one another could be determined.

It is important for the foundation of chemical theory to be able to substitute the mass for the volume of the hydrogen in these and similar cases, so as to be in a position to express the chemical equivalents of the metals in terms of hydrogen taken as the unit. In order to accomplish this, we must know the weight of a unit volume of hydrogen. This cannot be determined directly by any experimental process, which is accurate enough and at the same time sufficiently easy to be attempted by young students. But an equivalent result may be attained by making use of a form of experiment familiar in analytical chemistry, in which the loss of weight consequent upon the escape of gas from a light glass apparatus is determined. A plan has already been described by which the hydrogen expelled from an acid by a weighed quantity of metallic magnesium can be collected and measured ; by the following device the *weight* of this hydrogen can be determined.[1]

The apparatus consists of a wide-mouthed weighing bottle of thin blown glass. Through a

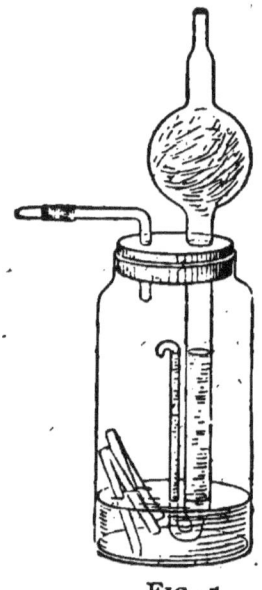

FIG. 5

rubber stopper pass two tubes ; one as shown bent at right angles and closed by a light rubber cap; the other shaped as shown in the figure, the bulb being filled with loose fibrous asbestos. About

---

[1] The suggestion that this can be accomplished was derived from Professor J. Emerson Reynolds's 'Experimental Chemistry for Junior Students' (Longmans), a book which deserves the attention of all teachers.

Chemical Equivalents

5 c.c. of water are placed in the bottle, together with the weighed quantity of metal. The bulb tube is dipped into a beaker of strong sulphuric acid till the lower part is filled, as shown in the sketch. The outside of this tube is then washed free from sulphuric acid, and the stopper bearing the tube fixed in its place. The whole apparatus is then placed on the pan of the balance, and weighed carefully, the weight being noted to milligrams. The whole apparatus should not weigh more than 30 to 40 grams. The weight having been recorded, a small piece of rubber tubing is attached to the open end of the bulb tube, and by blowing gently a few drops of sulphuric acid may be expelled into the water, in which the metal is immersed. Hydrogen is immediately evolved, and, escaping through the tube containing the oil of vitriol, is dried in its exit. When the whole of the metal has been dissolved, the cap is removed from the end of the right-angled tube, and air is gently sucked through the rubber attached to the bulb tube. The residual hydrogen is thus replaced by air, and, after a few minutes, the apparatus is ready to be weighed.

The following are the results of an experiment:

Magnesium taken . . . . . ·12 gr.
Weight of apparatus before solution of metal, 34·469 gr.
      „   „   after   „   „   34·459 gr.
                        Loss of weight   ·010 gr.

D

Thus magnesium expels from dilute acid $\frac{1}{12}$ of its weight of hydrogen, or 12 parts of magnesium replace 1 part by weight of hydrogen. Now, having shown by previous experiments that 1 gram of magnesium produces from dilute sulphuric acid about 930 c.c. of hydrogen and $\frac{1}{12}$ of a gram is ·0833, this is the weight of 930 c.c. of the gas: 1000 c.c. of hydrogen would, therefore, weigh ·0895, which is near enough to the figure usually given, viz. ·0896 gram.

The determination of the composition of water by weight is another experiment of the same order which may be done by pupils at about the same stage.

For this experiment we require a Kipp or other apparatus for generating a stream of hydrogen free from air, a couple of tubes containing pumice stone wetted with strong sulphuric acid in order to dry the gas, a tube containing pure dry oxide of copper which can be weighed, and an apparatus also light enough to be weighed, in which the water formed can be collected. The shape of a convenient apparatus for gathering the whole of the water without loss by vaporisation is shown in fig. 6, where *a* represents one of the tubes by which the stream of hydrogen is rendered dry; *b* is a bulb tube of hard glass containing pure dry black oxide of copper; the end of this tube is turned down at a right angle, and the extremity cut off obliquely. This tube is suspended by means of a thin piece of iron or, better, platinum wire from any suitable support, and by the same wire it can be attached to the beam of the balance when it has to be weighed. The thistle funnel has a bulb on the stem, the lower end of which dips through a cork into some oil of vitriol contained in a short test-tube. There is a short exit

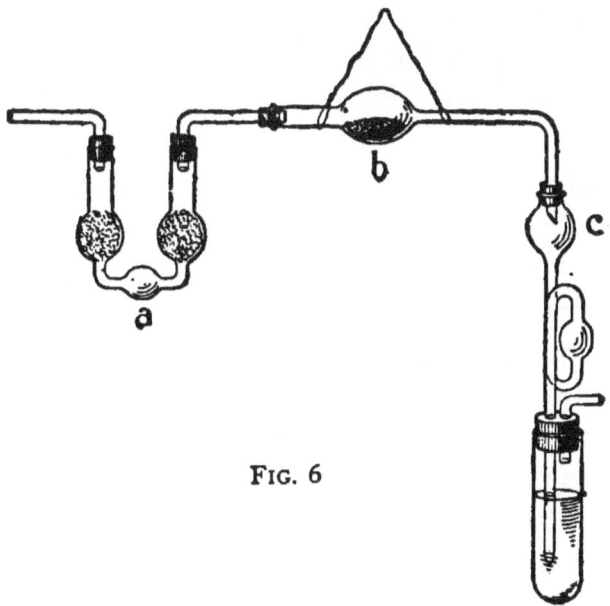

FIG. 6

tube from the cork for the escape of the excess of hydrogen. The whole of c can be hung on the balance by means of a wire loop. For the success of the experiment, the copper oxide should be freshly ignited, so as to ensure its dryness and freedom from nitrate or other impurity, and the stream of hydrogen should be carried through the whole series of tubes at a moderate speed for 5 to 10 minutes before the copper oxide is heated. b and c are to be weighed as accurately as possible at the temperature of the air before and after the reduction of the oxide and formation of the water. The gain of weight in c (water) should be to the loss of weight in b (oxygen) in the ratio of 9 to 8.

The fact that oxygen combines with twice its volume of hydrogen was observed by Cavendish in the middle of the last century, but it was Gay-Lussac[1] who established the generality of the principle that gaseous combinations are formed by the union of their constituents in very simple volumetric proportions. Thus he observed that

100 vols. of oxygen combine with 200 vols. of hydrogen.

100 vols. of muriatic acid combine with 100 vols. of ammonia.

100 vols. of carbon dioxide combine with 200 vols. of ammonia.

He also drew attention to the established composition of ammonia, which consists of 100 vols. of nitrogen to 300 vols. of hydrogen; the composition of sulphuric anhydride, which is formed of 100 vols. of sulphur dioxide united to 50 vols. of oxygen; of carbonic anhydride, which is composed of 100 vols. of carbonic oxide and 50 vols. of oxygen, and to the composition of the oxides of nitrogen. From all these facts Gay-Lussac drew conclusions which supplied very important arguments in favour of

[1] See 'Alembic Club Reprints,' No. 4, p. 15.

Dalton's Atomic Theory, at that time still under discussion, and far from being the generally accepted basis of chemical theory it is at the present day.

The experimental demonstration of the volumetric composition of hydrogen chloride, water, and ammonia is a matter of fundamental importance, but is best exhibited by the more experienced hand of the teacher. The methods generally in use for this purpose were for the most part devised by Hofmann, whose 'Introduction to Modern Chemistry,' though published a quarter of a century ago, is still delightful and instructive reading. As all these processes are described in the principal text-books, and particularly in detail in Mr. Newth's ' Chemical Lecture Experiments ' (Longmans), it is unnecessary to repeat the account of them in this place. One or two remarks, however, are necessary to indicate what each experiment serves to establish.

Hydrogen chloride.—It is usual to analyse hydrogen chloride gas by means of sodium amalgam. It must, however, be noted that this experiment merely proves that 2 vols. of the gas contain 1 vol. of hydrogen ; it gives no information as to the volume of the chlorine with which this 1 vol. of hydrogen is united.

The volume of the chlorine can only be shown by collecting the gas which results from electrolysis of the strong aqueous solution, and showing that it contains half its volume of chlorine and half its volume of hydrogen. The absorption of the chlorine from the mixed gas is best effected by means of concentrated solution of potassium iodide. Not only is half the gas absorbed, but the liberated iodine serves to demonstrate the nature of the gas.

Water.—As to the composition of water, it must be observed that the process of electrolysis so commonly resorted to proves only the ratio of the hydrogen to the oxygen, and is therefore incomplete as a demonstration of the composition of water. To show the volumetric relation of water gas to its constituents, the process must be synthetical—that is, the method must be used which consists in uniting 2 vols. of hydrogen with 1 vol. of oxygen at a temperature above 100° C.

Ammonia.—The ratio of the volume of hydrogen to that of nitrogen in ammonia is shown by the chlorine process; but the proof that 3 vols. of hydrogen and 1 vol. of nitrogen are contained in 2 vols. of ammonia is obtained only by decomposing a measured quantity of ammonia by heat, and subsequently analysing the resulting mixture of hydrogen and nitrogen. This is usually accomplished by sparking the gas till its volume is doubled, and then heating in the mixed gas a particle of cupric oxide, which converts the hydrogen into water.

The composition of ammonia cannot be conveniently demonstrated by exploding the gas with oxygen; for combustion with excess of oxygen leads to the production of so considerable a quantity of oxides of nitrogen as to render the process worthless for analytical purposes.

From the experiments just referred to we learn that equal volumes of chlorine, oxygen, and nitrogen combine respectively with one, two, and three volumes of hydrogen. Hence, *without any assumptions* as to atoms or molecules, we may express the composition of these gases by the formulæ HCl, H_2O, H_3N, in which the symbols H, Cl, O, N represent unit volumes of the constituents.

We know also that sodium and other metals acting upon these gases expel from them part of the hydrogen they contain, and take the place of the hydrogen so removed, the amount of hydrogen replaced being one-half in the case of water, and one-third in the case of ammonia, while from hydrogen chloride the whole of the hydrogen is removed at once. These facts are recorded in the formulæ.

In the earlier chapters great stress was laid upon the importance of cultivating the powers of observation; but, to make any progress with the theory of chemistry, it is equally necessary to practise the careful consideration and selection of evidence, and to form habits of reasoning thoughtfully from facts previously established. The confusion so often existing in the minds of students is, in general, due partly to the use of unfamiliar language, partly to want of experience in observation and in reasoning. This want of experience of phenomena and of the nature of scientific evidence renders the condition of the student-mind quite comparable with that of the early investigators. Some knowledge of the history of the development of modern chemistry from the time of Boyle onwards is therefore very valuable to the teacher. Frequent reference to a good English dictionary even by the teacher is also a practice which cannot be too strongly commended. The statement by a pupil that 'metals are hard and brittle, and are also malleable and ductile,' is a proof that the wholesome custom of explaining the meaning of words must have been neglected by that boy's teacher. Endless examples of the same defects are noticeable not only in the papers sent in by candidates at the examinations of the Science and Art Department, but at the Matriculation of the London University and other examinations of junior students within the experience of the author.

In the matter of selecting evidence great mistakes are commonly made, in many cases arising from mere indolent habit of mind or positive inability to think. A question recently set requiring evidence of the presence of oxygen in nitrous and nitric oxides was generally answered by the statement that 'they support combustion, therefore they contain oxygen.' Here 'support combustion' may be assumed to refer to the combustion of a

Chemical Equivalents 41

candle. In the first place, therefore, it is not true that they both support combustion; but this answer also ignores the fact that there are many known examples of burning where there is no oxygen. The proof required is as follows: Pass the gas over heated copper, then subject the hot copper oxide to a stream of hydrogen, and show the production of water. To make the proof complete, a knowledge of the composition of water must be assumed, and the water must be fully identified by observation of its physical as well as chemical characters.

A similar case occurs in the comparison of the gas obtained from chalk by heat or acids with that which is produced by combustion of charcoal, by fermentation of sugar or in animal respiration. It is not sufficient to point to the white precipitate with lime water as evidence of identity. The *probability* that they are the same is increased by each additional character—taste and odour, solubility, density, chemical character—which is found to be alike in both, but *certainty* is not attained till *all* available characters have been examined.

It may not be within the power of young students to make an exhaustive investigation of this kind, but the teacher should be careful to show that there are different stages or degrees of probability leading up to proof.

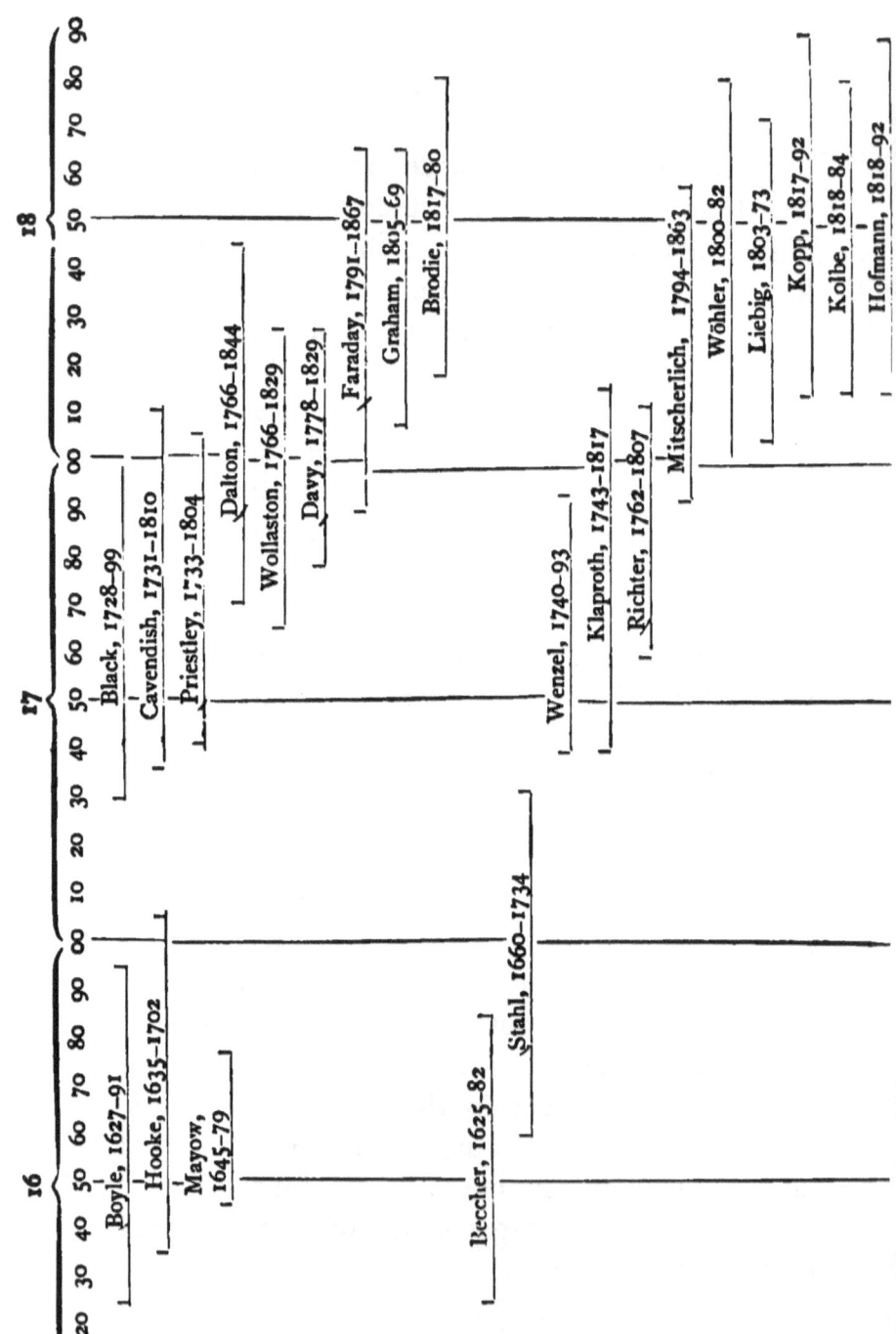

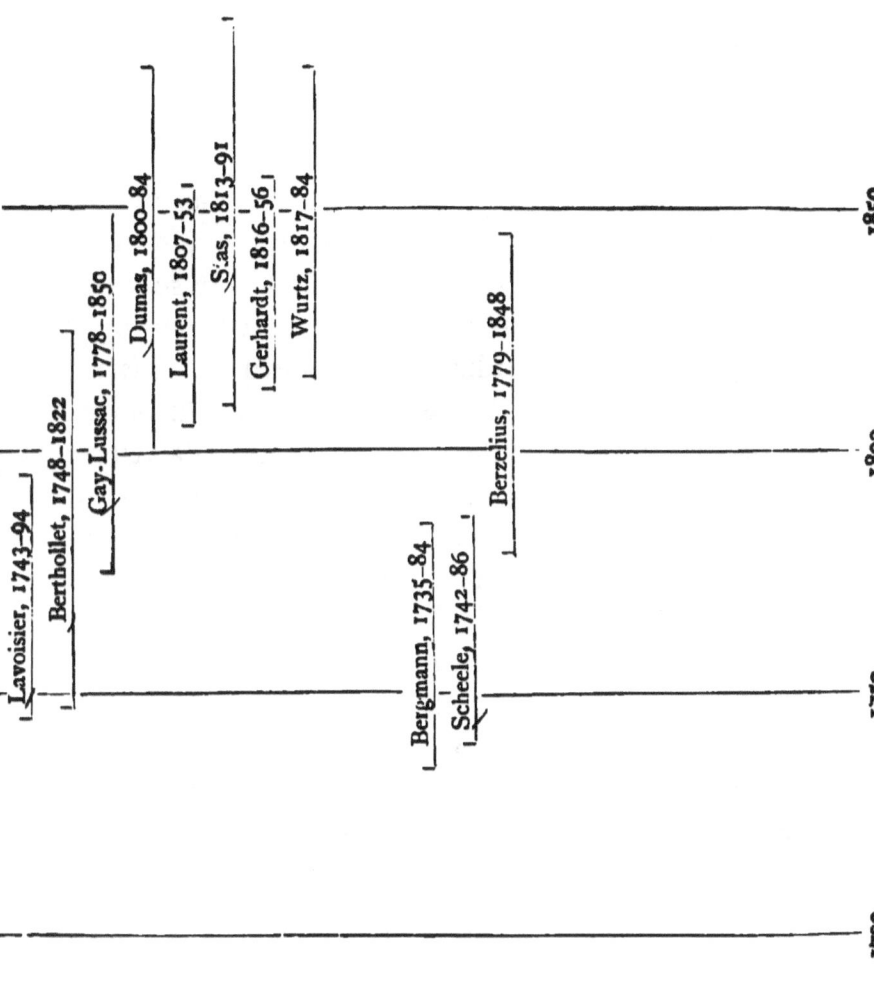

Note to Chapter IV.

The preceding chronological table was drawn up by the author some years ago, and it has been found very useful in his own teaching. It supplies the names of those chemists of the past who may be regarded as founders of modern chemistry, when they lived, and the length of each life, so as to show at a glance who were contemporary at any period. The table only includes the names of deceased chemists who have substantially advanced the theory of chemistry by their discoveries or by their writings. A crowd of well-known names has necessarily been omitted, as representing contributors to the detail rather than to the fundamental principles of the science, and in a few cases it may be a question whether the names in the list are all entitled to a place there. Klaproth, for example, did a great deal of work relating to the details of mineral analysis, and his name is admitted, chiefly on the ground that improvements in the art of analysis lent important aid to the progress of the science.

A general survey of the work of the majority of the chemists whose names appear in the table is provided in a compact form in Mr. Pattison Muir's volume entitled 'Heroes of Science— Chemists' (Society for Promoting Christian Knowledge, 1883).

There is also Watts's translation of the ' History

of Chemical Theory,' by the late Professor Wurtz (Macmillan), which brings the story down to 1869, now more than a quarter of a century ago.

Dr. Thorpe's charming volume of 'Essays in Historical Chemistry' (Macmillan) gives a critical review of the work and an account of the life of many, but not all, of these founders of Chemical Science.

CHAPTER V

MOLECULAR WEIGHTS AND FORMULÆ

THE establishment of Gay-Lussac's 'Law of Volumes' was destined to lead to consequences of the utmost importance for theoretical chemistry. In 1811 Avogadro published his discussion[1] of the constitution of gases upon the basis of Gay-Lussac's law, and although it attracted comparatively little notice at the time, what is now known as the law of Avogadro became half a century later the foundation of the modern system of molecular and atomic weights. The law of Avogadro states that equal volumes of different gases under the same conditions of temperature and pressure contain the same number of molecules. This statement is sometimes altered by students, and even by teachers, into 'gaseous molecules have the same or equal volumes.' This is wrong, because we know next to nothing about the dimensions of single molecules : all that we can

---

[1] 'Essay on a Manner of Determining the relative Masses of the Elementary Molecules of Bodies and the Proportions in which they enter into these Compounds.' See 'Alembic Club Reprints,' No. 4, p. 28.

Molecular Weights and Formulæ 47

say is that molecules of all gases require on the average equal spaces to move about in.

The density of a gas or vapour is the mass of unit volume, taking some gas, preferably the lightest of all, as the standard. So that when we say the density of oxygen is 16, of nitrogen 14, of carbonic anhydride 22, and so on, we mean that equal volumes of these gases weigh 16, 14, 22, &c. units when the weight of the same volume of hydrogen is 1 unit. Now it can be shown that the molecule of hydrogen is divisible into two equal parts. The hypothesis of the divisibility of many molecules is discussed very clearly in Avogadro's essay, to which reference has been made. The argument may run in this way: Equal volumes of hydrogen and chlorine interact to form hydrogen chloride, the volume of which is equal to the volume of the hydrogen added to that of the chlorine. If Avogadro's hypothesis is true, as we assume it to be, this process may be expressed by saying that a molecule of hydrogen reacting with a molecule of chlorine gives two molecules of hydrogen chloride. Hence each molecule of hydrogen chloride must contain half the hydrogen contained in a molecule of hydrogen, and half the chlorine contained in a molecule of that gas. Hence the molecules of hydrogen and of chlorine are respectively divisible into two equal parts, and the molecular weight of either gas may be conveniently taken to be the weight of two unit volumes. The determination of the density of a gas or vapour, therefore, leads directly to the molecular weight of the gas, whether elementary or compound.

The experimental methods of determining the densities of gases used by such great experimenters as Regnault and Lord Rayleigh are exceedingly simple in principle; but attainment to great accuracy requires much skill and experience, and attention to many details. But, though the method cannot be applied by beginners to the case of hydrogen and the lighter gases, it may be used even by young students with considerable success when

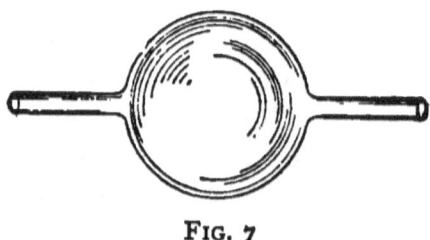

Fig. 7

the density of the gas is so great that the experimental errors of weighing and those due to impurity in the gas stand in small proportion relatively to the whole mass to be weighed.

Two globes are taken of the form shown in the figure. They may be purchased in the shape of large light pipettes, the stems of which may be cut off four or five cm. from the globe and fitted with light rubber caps. The globes should be so chosen as to be as nearly as possible of the same capacity—viz. 250 to 300 c.c. They should be hung by thin wires at the opposite ends of the beam of

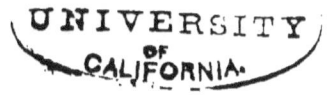

the balance without removing the pans, and the lighter one supplemented by a small piece of glass tubing filed down so as to supply an exact counterpoise. One of the globes which is to hold the gas must have its capacity determined once for all by filling with water and running it out into a measuring vessel. The globe is then dried and filled with the gas by displacement of air, the globe being held with the ends vertical, while the gas, if heavier than air, is admitted below.

The ends of the tubes are then closed again with the caps. On replacing the globe upon the balance, weights will have to be added to the opposite pan to restore equilibrium. These weights cannot be taken as a measure of the mass of the gas, as in the case of solids or liquids, inasmuch as the volume of air displaced, by the gas in this case, has an effect too great to be ignored. The weight of this displaced air must be calculated and added to the weights used in order to arrive at the weight of the gas in the globe. The details of an experiment will make the matter clear :

Capacity of globe 235 c.c.

Weight of 1 c.c. of air at normal temperature and pressure = ·0012937 gram. At 17°, the temperature of the laboratory, the weight of 1 c.c. of air is ·00122 *very nearly*. 235 c.c. therefore weigh ·286 gram. Globe filled with carbon dioxide required for counterpoise ·13 gram. Then ·286 + ·13 = ·416 is the weight of 235 c.c. of the gas.

The density is therefore $\frac{\cdot 416}{\cdot 286} = 1 \cdot 45$.

Calculated density 1·52 (air = 1).

The determination of the vapour density of various volatile solids or liquids can be so easily accomplished by Victor Meyer's air-expulsion method, and the calculations involved are so simple, that even young students may successfully practise the operation. It is not necessary to give an account of the process in this place, as it is described with full detail in many text-books. Ether (b.p. 35°) and chloroform (b.p. 61°) may be used as examples of substances which volatilise readily and completely in boiling water; alcohol (b.p. 78°) and benzene (b.p. 81°) do not give results so good, unless the water is saturated with common salt, which brings up the boiling point to 110°. The employment of higher temperatures requires rather more skill than beginners usually possess, but a bath of glycerine or oil, or any of the organic liquids of constant boiling point mentioned in the books, may be used when the pupil has had sufficient practice.

A knowledge of molecular weights provides one very important method by which atomic weights may be chosen. The atomic theory explains chemical phenomena by the assumption that combination results from the close approximation and connection of minute masses of elementary matter called 'atoms,' because they are believed to be

Molecular Weights and Formulæ 51

indivisible by the forces which operate in chemical changes. Chemical decomposition in like manner is the result of the separation of the atoms forming a compound and their rearrangement into new combinations. The hypothesis of atoms, however it may have originated and developed in Dalton's mind,[1] is firmly established by the facts which are embodied in the so-called laws of chemical combination. It is a fact that (1) elements unite in definite proportions; that (2), if two elements combine in several proportions, these proportions bear a simple relation to one another; that (3), if two elements combine with a third, the proportions in which they unite with this third element are the same proportions or simple multiples of the same proportions in which they combine together, should combination between them ever occur; and, lastly, that (4) when gases unite together, the volumes so uniting are represented by very simple proportions.

If now we accept the doctrine of Avogadro, the molecular weight is proportional to the specific gravity of the substance in vapour. For example, the molecule of carbon dioxide is 22 times heavier than the molecule of hydrogen, since their specific gravities are   22 : 1.

But for reasons already given, the molecule of hydrogen is believed to be divisible into two equal

---

[1] There seems to be some conflict of evidence on this point. See 'Life of Dalton,' by Sir H. E. Roscoe, 'Century' Series; also 'A New View of the Origin of Dalton's Atomic Theory,' by H. E. Roscoe and A Harden.

parts, or, in other words, consists of two atoms. Hence, to avoid fractions, the relative molecular weights are represented as

$$44 : 2.$$

Taking two volumes of hydrogen as the standard volume for comparison of molecular quantities of all gases, we have a means of fixing atomic weights.

For since by hypothesis the atom of an element is an indivisible quantity, it is evident that, if we measure off molecular proportions of all known volatile compounds containing that element, the smallest quantity of the element in question ever observed in such molecular proportion must be assumed to be one atom, until and unless some other compound is afterwards found containing a smaller quantity. This may be illustrated by the case of oxygen, as shown in the following table :

	Vapour density, or weight of 1 volume of vapour	Weight of 2 volumes of vapour	Weight of oxygen in 2 volumes of vapour
Water	9	18	16
Carbonic oxide	14	28	16
Carbonic anhydride	22	44	32
Nitrous oxide	22	44	16
Nitric oxide	15	30	16
Sulphurous anhydride	32	64	32
Sulphuric anhydride	40	80	48
Alcohol	23	46	16
Aldehyd	22	44	16
Acetic acid	30	60	32

As the smallest proportion of oxygen ever

Molecular Weights and Formulæ 53

found in two volumes of the vapour of any of its compounds is 16 units of weight, 16 is the value assigned to the atomic weight.

One word of caution is necessary in consequence of the confusion of phraseology still prevalent, even in the writings of chemical experts, in reference to atomic weights. In some of the principal chemical periodicals we find papers the titles of which announce that they refer to the 'atomic weights' of various elements, as, for example, in recent years the atomic weight of oxygen, of chromium, of titanium, of silicon, of tellurium, &c. In nearly all these cases the inquiry relates not to the determination of the atomic weight, but of the *combining proportion* or *equivalent* of the element in question. In fact, it may be stated as a general rule that two distinct operations are required for the determination of the *exact* value of the atomic weight of any element, and that the second of these is commonly assumed or ignored. First, the combining ratio, or equivalent, is determined with the utmost possible accuracy by the analysis or synthesis of some one or more of its compounds; and, secondly, the equivalent is converted into the atomic weight by multiplying this experimental number by some factor—1, 2, 3, 4, or more—which is chosen by appeal to some other wholly different criterion, such as that just referred to in the case of oxygen. The atomic weight of oxygen, in fact, is not 16 exactly but some number closely approaching 16 which is determined by analysis of water and other oxides.

Other examples and other methods are given in the next chapter.

CHAPTER VI

ATOMIC WEIGHTS—CLASSIFICATION

THE calculation of the atomic weight of an element must in all cases be preceded by the determination of the quantity of that element which combines with or replaces one unit weight of hydrogen. This is the equivalent. The operations which follow are of different kinds, according to the character of the element concerned, the most important being the following:

I. The *vapour density* method, described in the last chapter. It is a mistake to assume, as is sometimes done, that the vapour densities of the elements themselves serve as a guide to their atomic weights; for though it is true that the vapour densities and atomic weights are identical in a few cases, this affords no evidence towards the establishment of the atomic weight without additional information, and beside these are nearly as many other elements (Hg, Zn, Cd, As, P) whose vapour densities do not coincide with their atomic weights.

II. Appeal to the *Law of Dulong and Petit*

relating to *Specific Heat*. Care is necessary in expressing this law; it relates only to the elements in the *solid* state, and does not hold under ordinary conditions for the solids, boron, carbon, silicon, and beryllium.

The law may be stated in various ways: The specific heat of a solid element (with the named exceptions) is inversely as the atomic weight; or, the specific heat multiplied by the atomic weight is a constant (about 6·4); or the atomic weight is the quotient obtained by dividing 6·4 by the specific heat.

The application of the law consists in finding what multiple of the equivalent must be taken to comply with the formula

$$n \text{ Equiv.} \times \text{Sp. H.} = 6\cdot4 \text{ (approx.)},$$

the problem being to find the value of n. Now n must always be a whole number, because the atomic weight must always be identical with the equivalent or some multiple of it, according to the valency of the element. The atomic weight obviously cannot be deduced directly from the value of the specific heat, because no process is known by which the specific heat can be determined with the same degree of accuracy as the equivalent. This should be explained carefully, for students, even at the Honours stage, are too often impressed with the idea that the atomic weight can be directly settled by this process alone,

Atomic Weights—Classification 57

and seem to be ignorant that the specific heat can only be used by way of control.

III. Atomic weights can be deduced from equivalents by appeal to the *Periodic Law*.

The statement of the general principle may run thus: when the elements are ranged in a series in the order of the numerical value of their atomic weights, there is observed a revival of the same or closely similar characters *periodically*, and usually at every eighth term. This principle is usually displayed in the following or some similar table.

The utility of this table in the settlement of atomic weights depends upon the credence which is given to it. So many cases of doubtful atomic weights have, however, been submitted to this criterion, with results always tending to confirm the law, that it is now generally accepted. An example or two of its application will suffice to make it clear.

The element tellurium has an atomic weight which, deduced from different experimental estimations of its equivalent, probably lies somewhere between 127 and 128.

Tellurium is undoubtedly related to selenion in the same way that selenion is related to sulphur. Hence tellurium ought to stand in the table vertically under selenion; it cannot do so unless a number is adopted less than 127, and therefore differing by several units from the experimental result. In this case the experimental value has been set aside by the

58 Teaching of Elementary Chemistry

	Monads	Diads	Triads or Quasi-Triads	Tetrads	Triads or Pentads	Diads or Hexads	Monads or Triads	Valency various
1	Li7	Be9·1	B11	C12	N14	O16	F19	
2	Na23	Mg24	Al27	Si28	P31	S32	Cl35·5	
3	K39	Ca40	Sc44	Ti48	V51·4	Cr52	?	{ Mn55 { Co59 Cu63·3 Fe56 { Ni58
4	?	Zn65	Ga69	Ge72	As75	Se79	Br80	
5	Rb85·5	Sr87·6	Y89	Zr89·6	Nb94	Mo95·7	?	{ Ru104 Pd106 Ag108 Ro104
6	?	Cd112	In114	Sn118	Sb120	Te125?	I127	
7	Cs133	Ba137	{ Ce141 La139	?	?	?	?	
8	?	?	?{ Pr144 Sm150		?	?	?	
9	?	?	{ Er166 Yb173	?	Ta182	W184	?	{ Ir193 Pt195 Au197 Os199
10	?	Hg200	Tl204	Pb207	Bi208	?	?	
11	?	?	?	Th234	?	U239	?	

Atomic Weights—Classification 59

majority of chemists in favour of the theoretical, because it is thought to be more likely that an undetected source of error has attended the experiments than that the statement of the law should be at fault.

The element beryllium is a metal which was formerly supposed, chiefly on the evidence of the specific heat ($\cdot$445 at 10° to 100°), to have the atomic weight 13·5, but it was observed that there is no place in the scheme for an element having the combination of properties exhibited by beryllium associated with such a value for the atomic weight. Then it was discovered that the specific heat of beryllium is anomalous in the same sense as the specific heats of boron, carbon, and silicon especially, and, in a minor degree, of several other elements of comparatively low atomic weight. Redeterminations of the specific heat of beryllium at higher temperatures ($\cdot$62 at 400° to 500°), and applications of these values in the formula expressing the law of Dulong and Petit, led to the value 9·1 for the atomic weight, and this is now universally adopted. This value is confirmed by the vapour density of the chloride. Hence beryllium is now recognised as the first term of the series to which magnesium belongs.

When, therefore, the equivalent of an element is known, and a question is proposed as to its atomic weight, the answer may be obtained not only by referring to the specific heat, but by recalling the chief properties of the element, and then considering whether, having such properties, it can possibly occupy the position which would be assigned to it if the equivalent is accepted as equal to the atomic weight. Take the case of magnesium, the equivalent of which is 12·1. Magnesium is a light metal, forming one salifiable oxide which is insoluble in water; it forms a sulphate which crystallises in prisms containing water of crystallisation, and having the same form as sulphate of zinc. On referring to the table we see that there is no place between carbon and nitrogen available for any element, and that there is no element in any of the columns headed by boron, carbon, or nitrogen which exhibits properties similar to those of magnesium. But the isomorphism of the sulphate with sulphate of zinc at once gives the clue, and we see that the equivalent must be doubled in order to supply the value of the atomic weight. Magnesium with the value 24·2 or thereabouts then naturally falls into place as the first term of the series of which zinc and cadmium are the succeeding members.

The confidence placed in this table of the elements is based upon the belief that the properties of the elements are intimately dependent upon their atomic weights, that the number of elements is limited, and that their atomic weights can have only certain values. All experience in the past certainly tends to support this view. Thus in 1871 the places in the table now occupied by gallium, germanium, and scandium were vacant. These three elements have since been recognised with the properties attributed to them from a consideration of their position in the table.[1]

There is a strong analogy between this manner of displaying the connection between the elements and their atomic weights and the process of grouping the compounds of carbon into homologous series. Fifty years ago a considerable number of hydrocarbons, alcohols, aldehyds, bases &c. were known, but for the most part they were known only as individual substances without recognisable relations to one another. Then it was remarked, first by Schiel, and afterwards by Dumas, that the radicles of the alcohol and of the fatty acids exhibited a regularity of composition, and that the properties of the substances themselves which formed the series were only gradually modified in passing from term to term. Hence wood spirit, common alcohol, and fusel oil were found to be related to another, and even such apparently dissimilar substances as formic acid and stearic acid were recognised as terms of the same 'homologous' series.

The arrangement of the elements themselves into groups of closely related members of the same type, with properties gradually modified through successive terms, was a discovery of the same order,

[1] For a fuller exposition, see Mendeleeff's 'Principles of Chemistry,' vol. ii.

and since, as pointed out by Dumas, the values of the atomic weights exhibit the same kind of relations to one another as the atomic weights of the radicles in successive terms of a homologous series, the analogy is very remarkable. A single example will suffice:

($a = 7$ $d = 16$)		($a = 15$ $d = 14$)	
	Atomic weight.		Atomic weight.
Lithium	a	Methyl	a
Sodium	a + d	Ethyl	a + d
Potassium	a + 2d	Propyl	a + 2d

Now the several short series of elements when placed in parallel columns, as in the table given above, stand towards another in much the same position as heterologous series of carbon compounds; except, of course, that while the latter can be transformed into one another, the former cannot.

The periodic law, then, is based upon and includes the results of the earlier attempts at classification of the elements. All classification is founded upon the recognition of resemblances and differences, upon finding similarity in the midst of diversity; but to be useful a system must be based upon a record of as many points of resemblance as possible. Classification, therefore, implies, first, the practical process of *observation*, and, secondly, the survey of many observations, with a view to finding resemblances, so constituting the logical process of induction. In looking for indications of relationships, some considerable experience is necessary in many cases in order to select rightly among the phenomena observed and to assign to each its share of importance, so as to escape from the delusions to which the observer is exposed who gives too much attention to differences of colour, of density, or of mechanical condition. It is probably difficult for a young student to recognise at first the close relation subsisting between such substances as chlorine, bromine, and iodine. That a green gas should have much connection with a black shining solid, or even with a red liquid, doubtless appears at first a difficult proposition; and the conception that these three substances form a family having common features with minor differences comes only after the characters of a homologous series have been realised. It is on this account unfortunate that so-called Organic Chemistry should be separated from Inorganic Chemistry so rigidly as it commonly is by teachers, text-books, and boards of examiners. The idea may, however, be gained by the study of a single series of hydrocarbons—namely, the paraffins—in which the relation of physical properties such as volatility, density, melting point, solubility, to molecular weight can be readily illustrated. As the business of scientific chemistry is so largely made up of the art of classification, one or two further examples may be given.

The word 'metal' has received many applications, and even lexicographers seem to be in doubt as to its origin, but in chemistry it possesses a technical signification which is generally recognised. Nevertheless it is remarkable how few chemical teachers seem to have a clear idea of assigning to it a definite connotation, to judge at least by the answers given by candidates at examinations. Few, perhaps, are to be found in the confusion of mind exhibited by a student who, at a recent examination, stated that 'metals differ from non-metals, both by ending in *um* and having a metallic lustre,' but the answers of the best show that they have not learnt the art of observing correctly and selecting judiciously.

Dr. Percy, in the Introduction to his great work on Metallurgy, says: 'The term metal, like the term acid, is rather conventional than strictly scientific.' On the other hand, Professor Roberts-Austen, in his 'Introduction to the Study of Metallurgy,' remarks that 'the physical aspects of metals are so pronounced as to render it difficult to abandon the old view that metals are sharply defined from other elements, and form a class by themselves.' The difficulty is to select characters which may be regarded as diagnostic, marking off the members of the class from other bodies which do not belong to it.

The properties of weight and lustre seem formerly to have made up the whole idea of a metal, and the difficulty of getting rid of established prejudice, and of seeing things as they are, is illustrated in an amusing way by the well-known story of Dr. Pearson, a friend of Sir Humphry Davy's, who, on being shown a specimen of the newly discovered potassium, and noticing its lustre, exclaimed, 'Why, of course it is a metal—how *heavy* it is!'

Obviously, then, high density is not a property exhibited by all metals, nor even by all the common metals now that magnesium and aluminium are sold in the shops. Metallic lustre again is a quality exhibited by many substances, such as graphite, galena, pyrites, iodine, which are evidently not metals.

The following may be regarded as the characters of the metals as a class when in a pure state. All are more or less malleable and ductile, they are relatively good conductors of heat and electricity, and they form oxides which, when not too rich in oxygen, are basylous and interact with acids. The metals which present this assemblage of characters also rarely form compounds with hydrogen, and such of these compounds as are known are non-volatile solids.

The elements known, for want of a better name, as non-metals, though very diverse in physical characters, agree in forming volatile compounds with hydrogen and oxides which are, for the most part, capable of giving rise to acids by uniting with the elements of water. Natural objects, however, do not admit of classification by the application of

F

any rigid system, and no definition can be accepted absolutely and without reserve. The teacher will, therefore, at the proper stage, point out there are many elements which stand in an intermediate position between metal and non-metal, exhibiting, like tellurium and antimony, some of the properties of both classes of elements.

The difficulty of arriving at an agreement concerning the application of terms is still more forcibly displayed in the case of the words *acid, base, salt*. Even at the present day each of these words is encumbered with a residue of ancient usage from which it is almost impossible to set it free. The word 'salt' is perhaps less difficult now than it was a few years ago; but what is a base is still a subject concerning which no two chemists seem to be agreed.

Common salt is a food stuff known to mankind from very early times. The characters originally recognised were probably first its taste, then perhaps its solubility in water. But sodium chloride is not the only saline substance obtainable from the earth; there is nitre found commonly in tropical countries and the soda, borax, magnesium sulphate, sodium sulphate, and other crystalline deposits on the shores of many lakes. These were doubtless regarded in early times as closely akin to common salt from their saline taste, solubility, and possibly also from their crystalline appearance. Such was probably the vague kind of notion prevailing almost

Atomic Weights—Classification

down to the time of Lavoisier. Then came the dualistic theory of salts. Having found that metals combine with oxygen, and that the products are usually capable of dissolving in acids, with production of salts, Lavoisier saw in these facts the basis of a series of definitions. A salt is composed of an acid united to a base: an acid is an oxide, usually of a non-metallic element; a base is the oxide of a metal. If these views had stood the test of experience, we should now be in possession of a system consistent with itself. As it is, this view of the constitution of salts has been overthrown, while its nomenclature has been retained. The difficulty about the use of the term *base* arises from the fact, that though we may still apply it to certain of the oxides of metals, these compounds do not unite as a whole with acids, for water is now known to be formed in every case and eliminated as a bye product. The essential components of the idea of a base were in Lavoisier's time (1) that it was an oxide, and (2) that it entered into combination with an acid, tending to neutralise it. The only substances which enter wholly into union with acids are ammonia and its derivatives: these are the only true *bases*, but they do not contain oxygen. It is not very creditable to the leaders among chemists that this state of confusion in a comparatively important part of chemical nomenclature, otherwise reasonable and consistent, should be allowed to continue.

As to 'acid,' everyone now admits that this term is included under 'salt;' and to the question 'what is an acid?' a complete and legitimate answer would be 'a salt of hydrogen.' Fortunately we now possess a criterion by which a salt can be defined, no matter whether it be represented by the now obsolete dualistic symbol, or by the unitary or by the more modern structural formula.

Sodium sulphate, for example, is $Na_2O.SO_3$ or Na_2SO_4 or $(NaO)_2SO_2$.

The question whether it is a salt is now resolved by another—Does it readily enter into double decomposition, and is it an electrolyte? By this test we can at once distinguish salts from other compounds which in *formula* resemble them; thus sodium chloride, NaCl, is a salt; carbon chloride, CCl_4, is not a salt.

On the other hand, if this criterion is accepted, metallic oxides and hydroxides, the compounds to which commonly the term *base* is applied, come under the same category, caustic potash being nearly as good an electrolyte as hydrochloric acid.

These subjects are still undergoing investigation, and chemists are far from being united in the interpretation of experimental results. The teacher will do well to study current literature, and consider carefully the facts for himself.

While, according to the experience of the

author, it is unwise to communicate unsettled questions to mere beginners, as tending only to confusion and adding to their difficulties, it is imperative that all the facts leading up to any generalisation should be laid without reserve before advanced students. They should be encouraged to judge of evidence and draw conclusions for themselves, being instructed, however, that, no matter how firmly established or complete a theory (=a view) or hypothesis (=supposition) may seem to be, it is never final, and is certain to be sooner or later absorbed into a more comprehensive system. One of the greatest evils of working wholly by text-books without personal contact with a teacher who is master of his subject is that the student thus taught invariably imbibes the idea that the subject is complete, rounded off, and finished, and he sees no room for further inquiry. This is not only disastrous from the educational point of view, but is a serious disadvantage to the student who learns chemistry for the sake of its practical technical applications.

One word in conclusion. Chemistry cannot be learnt by reading alone. There comes a stage in every student's career when much reading is very necessary, but this is only reached after long training of the eye and the hand. First of all learn to see things clearly, and from first to last let fact stand before all hypothesis.

APPENDIX

EXTRACTS FROM THE SYLLABUS ISSUED BY THE DEPARTMENT OF SCIENCE AND ART, 1895.

SUBJECT X.—INORGANIC CHEMISTRY, THEORETICAL.

FIRST STAGE OR ELEMENTARY COURSE.

Chemical distinguished from physical changes. Indestructibility of matter. Decomposible and undecomposible substances. Action of heated copper and mercury on air. Action of heated iron on steam. Gas formed by action of iron on steam passed over product of heating copper in air gives water. Action of metals on hydrochloric acid gas and upon solutions of hydrochloric acid and sulphuric acid. Production of hydrogen. Production of oxygen from red mercuric oxide. Formation of water by combustion of hydrogen in air or by explosion with oxygen in eudiometer. Identification of water by its physical and chemical characters. Mixtures distinguished from compounds. Laws of definite and multiple combination. Equivalents. Estimation of equivalents illustrated experimentally by deposition of such metals as silver or copper by zinc or magnesium and weighing. Physical properties of gases as distinguished from liquids and solids. Boyle's law. Law of expansion

Appendix

of gases by heat (without problems). Phenomena of gaseous diffusion. Molecules and atoms. Formation and properties of products of burning carbon in excess of air or oxygen. Comparison of gas so formed with gas obtained by treating chalk &c. with acid. Composition of the atmosphere and action of animal and vegetable life. Systematic study of following elements and compounds, their condition in nature, usual methods of isolation and chief properties : hydrogen, oxygen, nitrogen, chlorine, hydrogen chloride (bromine and iodine to be exhibited and compared with chlorine), sulphur, sulphur dioxide, sulphur trioxide, and sulphurous and sulphuric acids, and a few of their most familiar salts, hydrogen sulphide, nitric acid and chief nitrates, nitrous oxide (from ammonium nitrate), nitric oxide (from nitric acid by copper), ammonia, carbon and its two oxides, properties common to acids in general.

Symbolic notation. Nomenclature. Formulæ and equations. Calculation of quantities by weight from chemical equations (French system of weights). Division of elements broadly into metals and non-metals. General characters of metals. Basylous oxides and hydroxides, such as lime, caustic soda, zinc oxide, black oxide of copper, and litharge, and the interaction of these with acids to form salts and water.

ALTERNATIVE FIRST STAGE OR ELEMENTARY COURSE.

No substantial alteration has been made in this part of the Syllabus, but a few corrections have been introduced into the phraseology.

SECOND STAGE OR ADVANCED COURSE.

In addition to the subjects of the first syllabus of the elementary course, students presenting themselves for the advanced examination will be assumed to have received instruction in the following:

The experimental methods by which the composition of the following bodies has been accurately determined: water, atmospheric air, hydrochloric acid, ammonia, the gaseous oxides of nitrogen, the oxides of carbon, sulphuretted hydrogen.

Laws of gaseous combination of elements and compounds. Reduction of gaseous volumes to standard pressure and temperature. Calculation of quantities by volume and by weight.

Atoms and molecules, atomic and molecular weights. Law of Avogadro. Atomic and molecular formulæ and equations. Specific heat and atomic heat of the elements. Graphic or constitutional formulæ. Atomic value or valency of the elements. Phenomena of dissociation. Classification of the elements.

WATER.—Causes of permanent and temporary hardness in. Modes of softening. Suitability for domestic purposes. Determination of composition of water by weight and by volume.

HYDROGEN DIOXIDE.—Its preparation and properties.

OZONE.—Production, properties and constitution. Occurrence in nature.

ATMOSPHERIC AIR.—Mode of ascertaining exact composition of. Carbon dioxide, amount contained in, and determination of. Aqueous vapour, determination of.

CHLORINE.—Theory of bleaching. Composition and preparation of bleaching powder. Oxides and the following oxy-acids of chlorine, viz.: Hypochlorous, chloric and perchloric acids, preparation, properties, and composition of.

BROMINE, and hydrobromic and bromic acids.

Appendix

IODINE, and hydriodic, iodic, and periodic acids.

FLUORINE.—Hydrofluoric acid.

PHOSPHORUS, its sources and its allotropic modifications; its hydrides, chlorides, and oxides. Phosphoric acids and the phosphates. Hypophosphorous and phosphorous acids.

ARSENIC and its hydrides, chlorides, and oxides. Arsenious and arsenic acids. Arsenites and arsenates. Detection of arsenic.

ANTIMONY and BISMUTH.—Hydride, chloride and oxides of antimony, chloride, oxides and salts of bismuth to be studied chiefly with the object of showing the relations of these two elements to the members of the phosphorus group.

BORON, its occurrence and allotropic modifications. Boron trioxide. Boric acid.

SILICON and silica. Silicic acid. Silicon hydride, silicon fluoride. Names and formulæ of some of the more important silicates (mineral).

The chief properties of the following metals, and the composition and properties of their *more important* compounds: potassium, sodium, ammonium, silver, mercury, copper, zinc, cadmium, magnesium, barium, strontium, calcium, tin, gold, aluminium, platinum, lead, chromium, manganese, iron, cobalt, and nickel.

The processes of formation and crystallisation of the chief salts of these metals should be practically demonstrated, and specimens should be prepared by the students.

Manufacturing processes for the production of—

Oxygen.
Chlorine, bromine, iodine.
Bleaching powder.
Sulphuric acid.
Hydrochloric acid.
Nitric acid.
Sodium carbonate and caustic soda.

Lime.
Iron and steel.
Copper.
Lead.
Mercury.

Silver and gold.
Zinc.
Aluminium.
Sodium.

SUBJECT X. *p.*—INORGANIC CHEMISTRY, PRACTICAL.

FIRST STAGE OR ELEMENTARY COURSE.

The practical knowledge of the candidate will in this stage be tested both by a written and a practical examination.

I. The subjects of the written examination will include—

(*a*) The preparation of the elements and compounds enumerated in the elementary course of Subject X. (Theoretical), and the methods of experimentally demonstrating their properties.

(*b*) The principal reactions, both wet and dry, of the following: lead, copper, iron, zinc, calcium, potassium, ammonium, carbonates, nitrates, sulphates, chlorides.

These questions will, as much as possible, be so framed as to prevent answers being given by students who have obtained their information merely from books and oral instruction. Any student on whom it is intended to claim payments in this stage may be called on by the inspector of the Department, when visiting the Laboratory, to repeat some of the experiments which he has had the opportunity of witnessing.

The value of the answers will be greatly enhanced by the neatness and clearness of the sketches; provided always, that an accurate knowledge of the construction of the apparatus is exhibited.

Appendix

II. The practical examination may consist of exercises selected from any of the following divisions of experimental work :

A.—Testing of two powders neither of which will contain more than two metallic and one acid radicle from the above list. The powders will be soluble in water or in dilute acid.

The experiments made must be carefully described, and the analytical results clearly stated and confirmed by more than one reaction.

No notes, books, or analytical tables may be consulted during the examination.

B.—Carrying out experiments for which printed instructions will be given in the paper, e.g. the observation of the effects of heat or of water, or acids &c. upon the materials supplied. If a gas is evolved, the candidate may be required to determine what it is. [Such a gas will always be one of those referred to under Subject X., First Stage, and of which the properties have been demonstrated in the lectures.]

C.—Recognition of materials which have been exhibited upon the lecture table, and the properties of which have been demonstrated, e.g. sulphur, iodine, charcoal, manganese dioxide, potassium chlorate, nitre, sal-ammoniac, &c.

D.—Quantitative experiments of a simple kind, e.g. a metal or a salt may be supplied to be weighed, dissolved in acid, and the amount of evolved gas measured at the temperature of the Laboratory. Or the carbon dioxide in a carbonate may be estimated by solution in an acid, and observation of the loss of weight. Or determination of the weight of one metal, such as silver, which can be displaced by another, such as magnesium. Or experiments on the neutralisation of acids or alkalies.

SECOND STAGE OR ADVANCED COURSE.

The examination will consist of two parts:

I. A short written examination of about four questions, with the object of testing the candidate's knowledge of the theory of ordinary methods of qualitative analysis and the preparation of such bodies as are enumerated in Second Stage, Subject X.

II. A practical examination, in qualitative and very simple quantitative analysis. One or two substances for qualitative analysis, each containing not more than four salt radicles, positive or negative, selected from the following list: silver, lead, mercury, copper, bismuth, cadmium, tin, arsenic, antimony, iron, manganese, aluminium, chromium, zinc, cobalt, nickel, calcium, strontium, barium, magnesium, potassium, sodium, ammonium, oxides, hydroxides, chlorides, bromides, iodides, fluorides, sulphides, sulphites, sulphates, chromates, carbonates, phosphates, arsenates, borates, nitrates, nitrites, chlorates, permanganates.

There might, therefore, be four metals in the form of oxide, or three metals in the form of the same salt, or two simple salts or one metal in the form of three salts.

One substance may be given to be tested quantitatively by means of a previously prepared volumetric solution.

The remarks with respect to inspection made in the Elementary Stage apply also to this grade.

Three and a quarter hours will be allowed for the examination in practical analysis, and one hour for the written examination.

NOTE.—Candidates will not be allowed to communicate with one another during the examination, but the use of notes or text-books of analysis will be permitted.

Spottiswoode & Co. Printers, New-street Square, London.

A CLASSIFIED CATALOGUE

OF

SCIENTIFIC WORKS

PUBLISHED BY

MESSRS. LONGMANS, GREEN, & CO.

LONDON: 39 PATERNOSTER ROW, E.C.
NEW YORK: 91 & 93 FIFTH AVENUE.
BOMBAY: 32 HORNBY ROAD.

CONTENTS.

	PAGE		PAGE
Advanced Science Manuals	30	Mechanics	6
Agriculture and Gardening	26	Medicine and Surgery	19
Astronomy	14	Metallurgy	14
Biology	24	Mineralogy	14
Botany	25	Natural History	18
Building Construction	10	Navigation	14
Chemistry	2	Optics	8
Dynamics	6	Photography	8
Electricity	11	Physics	5
Elementary Science Manuals	30	Physiography	17
Engineering	12	Physiology	24
Evolution	18	*Proctor's (R. A.) Works*	15
Geology	17	Sound	8
Health and Hygiene	17	Statics	6
Heat	8	Steam, Oil, and Gas Engines	9
Hydrostatics	6	Strength of Materials	12
Light	8	Technology	17
London Science Class-Books	32	Telegraphy	12
Longmans' Civil Engineering Series	13	Telephone	12
		Text-Books of Science	28
Machine Drawing and Design	13	Thermodynamics	8
Magnetism	11	*Tyndall's (John) Works*	27
Manufactures	17	Workshop Appliances	14

CHEMISTRY.

***CORNISH.*—PRACTICAL PROOFS OF CHEMICAL LAWS:**
A Course of Experiments upon the Combining Proportions of the Chemical Elements. By VAUGHAN CORNISH, M.Sc., Associate of the Owens College, Manchester. Crown 8vo., 2s.

***CROOKES.*—SELECT METHODS IN CHEMICAL**
ANALYSIS, chiefly Inorganic. By WILLIAM CROOKES, F.R.S., etc. Third Edition, Rewritten and Enlarged. With 67 Woodcuts. 8vo., 21s. net.

***CROSS* AND *BEVAN.*—CELLULOSE:** an Outline of the Chemistry of the Structural Elements of Plants. With Reference to their Natural History and Industrial Uses. By CROSS and BEVAN (C. F. Cross, E. J. Bevan, and C. Beadle). With 14 Plates. Crown 8vo., 12s. net.

***FURNEAUX.*—ELEMENTARY CHEMISTRY,** Inorganic and Organic. By W. FURNEAUX, F.R.C.S., Lecturer on Chemistry, London School Board. With 65 Illustrations and 155 Experiments. Crown 8vo., 2s. 6d.

***HJELT.*—PRINCIPLES OF GENERAL ORGANIC CHE-**
MISTRY. By Professor E. HJELT, of Helsingfors. Translated from the German by J. BISHOP TINGLE, Ph.D., Assistant in the Laboratory of the Heriot Watt College, Edinburgh. Crown 8vo., 6s. 6d.

***JAGO.*—Works by W. JAGO, F.C.S., F.I.C.**

    INORGANIC CHEMISTRY, THEORETICAL AND PRACTICAL. With an Introduction to the Principles of Chemical Analysis, Inorganic and Organic. With 63 Woodcuts and numerous Questions and Exercises. Fcp. 8vo., 2s. 6d.

    AN INTRODUCTION TO PRACTICAL INORGANIC CHEMISTRY. Crown 8vo., 1s. 6d.

    INORGANIC CHEMISTRY, THEORETICAL AND PRACTICAL. A Manual for Students in Advanced Classes of the Science and Art Department. With Plate of Spectra and 78 Woodcuts. Crown 8vo., 4s. 6d.

***KOLBE.*—A SHORT TEXT-BOOK OF INORGANIC**
CHEMISTRY. By Dr. HERMANN KOLBE. Translated and Edited by T. S. HUMPIDGE, Ph.D. With 66 Illustrations. Crown 8vo., 8s. 6d.

***MENDELÉEFF.*—THE PRINCIPLES OF CHEMISTRY.**
By D. MENDELÉEFF, Professor of Chemistry in the University of St. Petersburg. Translated by GEORGE KAMENSKY, A.R.S.M., of the Imperial Mint, St. Petersburg, and Edited by A. J. GREENAWAY, F.I.C., Sub-Editor of the Journal of the Chemical Society. With 97 Illustrations. 2 vols. 8vo., 36s.

***MEYER.*—OUTLINES OF THEORETICAL CHEMISTRY.**
By LOTHAR MEYER, Professor of Chemistry in the University of Tübingen. Translated by Professors P. PHILLIPS BEDSON, D.Sc., and W. CARLETON WILLIAMS, B.Sc. 8vo., 9s.

***MILLER.*—INTRODUCTION TO THE STUDY OF IN-**
ORGANIC CHEMISTRY. By W. ALLEN MILLER, M.D., LL.D. With 71 Woodcuts. Fcp. 8vo., 3s. 6d.

CHEMISTRY—*Continued.*

NEWTH.—Works by G. S. NEWTH, F.I.C., F.C.S., Demonstrator in the Royal College of Science, London; Assistant Examiner in Chemistry, Science and Art Department.

CHEMICAL LECTURE EXPERIMENTS, NON-METALLIC ELEMENTS. With 230 Diagrams. Crown 8vo., 10s. 6d.

A TEXT-BOOK OF INORGANIC CHEMISTRY. With 146 Illustrations. Crown 8vo., 6s. 6d.

ELEMENTARY PRACTICAL CHEMISTRY: a Laboratory Manual for Use in Organised Science Schools. With 108 Illustrations and 254 Experiments. Crown 8vo., 2s. 6d.

OSTWALD.—SOLUTIONS. By W. OSTWALD, Professor of Chemistry in the University of Leipzig. Being the Fourth Book, with some additions, of the Second Edition of Oswald's 'Lehrbuch der allgemeinen Chemie'. Translated by M. M. PATTISON MUIR, Fellow of Gonville and Caius College, Cambridge. 8vo., 10s. 6d.

PAYEN.—INDUSTRIAL CHEMISTRY. A Manual for use in Technical Colleges and Schools, based upon a Translation of Stohmann and Engler's German Edition of PAYEN'S *Précis de Chimie Industrielle*. Edited by B. H. PAUL, Ph.D. With 698 Woodcuts. 8vo., 42s.

REYNOLDS.—EXPERIMENTAL CHEMISTRY FOR JUNIOR STUDENTS. By J. EMERSON REYNOLDS, M.D., F.R.S., Professor of Chemistry, University of Dublin; Examiner in Chemistry, University of London. Fcp. 8vo., with numerous Woodcuts.

Part I. Introductory. Fcp. 8vo., 1s. 6d.

Part II. Non-Metals, with an Appendix on Systematic Testing for Acids. Fcp. 8vo., 2s. 6d.

Part III. Metals, and Allied Bodies. Fcp. 8vo., 3s. 6d.

Part IV. Carbon Compounds. Fcp. 8vo., 4s.

SHENSTONE.—Works by W. A. SHENSTONE, Lecturer on Chemistry in Clifton College.

THE METHODS OF GLASS-BLOWING. For the use of Physical and Chemical Students. With 42 Illustrations. Crown 8vo., 1s. 6d.

A PRACTICAL INTRODUCTION TO CHEMISTRY. Intended to give a Practical acquaintance with the Elementary Facts and Principles of Chemistry. With 25 Illustrations. Crown 8vo., 2s.

CHEMISTRY—*Continued.*

THORPE.—A DICTIONARY OF APPLIED CHEMISTRY. By T. E. Thorpe, B.Sc. (Vict.), Ph.D., F.R.S., Treas. C.S., Professor of Chemistry in the Royal College of Science, South Kensington. Assisted by Eminent Contributors. 3 vols. 8vo. Vols. I. and II., 42*s.* each. Vol. III., 63*s.*

THORPE.—QUANTITATIVE CHEMICAL ANALYSIS. By T. E. Thorpe, Ph.D., F.R.S. With 88 Woodcuts. Fcp. 8vo., 4*s.* 6*d.*

THORPE AND MUIR.—QUALITATIVE CHEMICAL ANALYSIS AND LABORATORY PRACTICE. By T. E. Thorpe, Ph.D. D.Sc., F.R.S., and M. M. Pattison Muir, M.A. With Plate of Spectra and 57 Woodcuts. Fcp. 8vo., 3*s.* 6*d.*

TILDEN.—Works by WILLIAM A. TILDEN, D.Sc. London, F.R.S., Professor of Chemistry in the Royal College of Science, London.

    INTRODUCTION TO THE STUDY OF CHEMICAL PHILOSOPHY. The Principles of Theoretical and Systematic Chemistry. With 5 Woodcuts. With or without the ANSWERS of Problems. Fcp. 8vo., 4*s.* 6*d.*

    PRACTICAL CHEMISTRY. The principles of Qualitative Analysis. Fcp. 8vo., 1*s.* 6*d.*

    HINTS ON THE TEACHING OF ELEMENTARY CHEMISTRY IN SCHOOLS AND SCIENCE CLASSES. With 7 Illustrations. Crown 8vo., 2*s.*

WATTS' DICTIONARY OF CHEMISTRY. Revised and entirely Re-written by H. Forster Morley, M.A., D.Sc., Fellow of, and lately Assistant-Professor of Chemistry in University College, London; and M. M. Pattison Muir, M.A., F.R.S.E., Fellow, and Prælector in Chemistry, of Gonville and Caius College, Cambridge. Assisted by Eminent Contributors. 4 vols. 8vo. Vols. I. and II., 42*s.* each. Vol. III., 50*s.* Vol. IV., 63*s.*

WHITELEY.—Works by R. LLOYD WHITELEY, F.I.C., Principal of the Municipal Science School, West Bromwich.

    CHEMICAL CALCULATIONS. With Explanatory Notes, Problems and Answers, specially adapted for use in Colleges and Science Schools. With a Preface by Professor F. Clowes, D.Sc. (Lond.), F.I.C. Crown 8vo., 2*s.*

    ORGANIC CHEMISTRY: the Fatty Compounds. With 45 Illustrations. Crown 8vo., 3*s.* 6*d.*

PHYSICS, ETC.

EARL.—THE ELEMENTS OF LABORATORY WORK:
a Course of Natural Science. By A. G. EARL, M.A., F.C.S., late Scholar of Christ's College, Cambridge. With 57 Diagrams and numerous Exercises and Questions. Crown 8vo., 4s. 6d.

GANOT.—Works by PROFESSOR GANOT. Translated and Edited by E. ATKINSON, Ph.D., F.C.S.

ELEMENTARY TREATISE ON PHYSICS, Experimental and Applied. With 9 Coloured Plates and Maps, and 1028 Woodcuts, and Appendix of Problems and Examples with Answers. Crown 8vo., 15s.

NATURAL PHILOSOPHY FOR GENERAL READERS AND YOUNG PERSONS; With 7 Plates, 624 Woodcuts, and an Appendix of Questions. Crown 8vo., 7s. 6d.

GLAZEBROOK AND SHAW.—PRACTICAL PHYSICS. By R. T. GLAZEBROOK, M.A., F.R.S., and W. N. SHAW, M.A. With 134 Woodcuts. Fcp. 8vo., 7s. 6d.

GUTHRIE.—MOLECULAR PHYSICS AND SOUND. By F. GUTHRIE, Ph.D. With 91 Diagrams. Fcp. 8vo., 1s. 6d.

PHYSICAL AND ELECTRICAL ENGINEERING LABORATORY MANUALS.—Vol. 1.
HENDERSON.—ELEMENTARY PHYSICS. By JOHN HENDERSON, B.Sc. (Edin.), A.I.E.E., Lecturer in Physics, Manchester Municipal Technical School. Crown 8vo., 2s. 6d.

HELMHOLTZ.—POPULAR LECTURES ON SCIENTIFIC SUBJECTS. By HERMANN VON HELMHOLTZ. Translated by E. ATKINSON, Ph.D., F.C.S., formerly Professor of Experimental Science, Staff College. With 68 Illustrations. 2 vols., crown 8vo., 3s. 6d. each.

CONTENTS.—Vol. I.—The Relation of Natural Science to Science in General—Goethe's Scientific Researches—The Physiological Causes of Harmony in Music—Ice and Glaciers—The Interaction of the Natural Forces—The Recent Progress of the Theory of Vision—The Conservation of Force—The Aim and Progress of Physical Science.

CONTENTS.—Vol. II.—Gustav Magnus. In Memoriam—The Origin and Significance of Geometrical Axioms—The Relation of Optics to Painting—The Origin of the Planetary System—Thought in Medicine—Academic Freedom in German Universities—Hermann Von Helmholtz—An Autobiographical Sketch.

WATSON.—ELEMENTARY PRACTICAL PHYSICS: a Laboratory Manual for Use in Organised Science Schools. By W. WATSON, B.Sc. Demonstrator in Physics in the Royal College of Science, London; Assistant Examiner in Physics, Science and Art Department. With 119 Illustrations and 193 Exercises. Crown 8vo., 2s. 6d.

PHYSICS, ETC.—*Continued.*

WORTHINGTON.—A FIRST COURSE OF PHYSICAL LABORATORY PRACTICE. Containing 264 Experiments. By A. M. WORTHINGTON, M.A., F.R.S. With Illustrations. Crown 8vo., 4s. 6d.

WRIGHT.—ELEMENTARY PHYSICS. By MARK R. WRIGHT, Professor of Normal Education, Durham College of Science. With 242 Illustrations. Crown 8vo., 2s. 6d.

MECHANICS, DYNAMICS, STATICS, HYDROSTATICS, ETC.

BALL.—A CLASS-BOOK OF MECHANICS. By Sir R. S. BALL, LL.D. 89 Diagrams. Fcp. 8vo., 1s. 6d.

GELDARD.—STATICS AND DYNAMICS. By C. GELDARD, M.A., formerly Scholar of Trinity College, Cambridge. Crown 8vo., 5s.

GOODEVE.—Works by T. M. GOODEVE, M.A., formerly Professor of Mechanics at the Normal School of Science, and the Royal School of Mines.

    THE ELEMENTS OF MECHANISM. With 342 Woodcuts. Crown 8vo., 6s.

    PRINCIPLES OF MECHANICS. With 253 Woodcuts and numerous Examples. Crown 8vo., 6s.

    A MANUAL OF MECHANICS: an Elementary Text-Book for Students of Applied Mechanics. With 138 Illustrations and Diagrams, and 188 Examples taken from the Science Department Examination Papers, with Answers. Fcp. 8vo., 2s. 6d.

GRIEVE.—LESSONS IN ELEMENTARY MECHANICS. By W. H. GRIEVE, P.S.A., late Engineer, R.N., Science Demonstrator for the London School Board, etc.

    Stage I. With 165 Illustrations and a large number of Examples. Fcp. 8vo., 1s. 6d.

    Stage 2. With 122 Illustrations. Fcp. 8vo., 1s. 6d.

    Stage 3. With 103 Illustrations. Fcp. 8vo., 1s. 6d.

Scientific Works published by Longmans, Green, & Co. 7

___

MECHANICS, DYNAMICS, STATICS, HYDROSTATICS, ETC.—
Continued.

MAGNUS.—Works by SIR PHILIP MAGNUS, B.Sc., B.A.

    LESSONS IN ELEMENTARY MECHANICS. Introductory to the study of Physical Science. Designed for the Use of Schools, and of Candidates for the London Matriculation and other Examinations. With numerous Exercises, Examples, Examination Questions, and Solutions, etc., from 1870-1895. With Answers, and 131 Woodcuts. Fcp. 8vo., 3s. 6d.

        Key for the use of Teachers only, price 5s. 3½d.

    HYDROSTATICS AND PNEUMATICS. Fcp. 8vo., 1s. 6d.; or, with Answers, 2s. The Worked Solutions of the Problems, 2s.

ROBINSON.—ELEMENTS OF DYNAMICS (Kinetics and Statics). With numerous Exercises. A Text-book for Junior Students. By the Rev. J. L. ROBINSON, B.A. Crown 8vo., 6s.

SMITH.—Works by J. HAMBLIN SMITH, M.A.

    ELEMENTARY STATICS. Crown 8vo., 3s.

    ELEMENTARY HYDROSTATICS. Crown 8vo., 3s.

    KEY TO STATICS AND HYDROSTATICS. Crown 8vo., 6s.

TATE.—EXERCISES ON MECHANICS AND NATURAL PHILOSOPHY. By THOMAS TATE, F.R.A.S. Fcp. 8vo., 2s. Key, 3s. 6d.

TAYLOR.—Works by J. E. TAYLOR, M.A., B.Sc. (Lond.), Head Master of the Central Higher Grade and Science School, Sheffield.

    THEORETICAL MECHANICS, including Hydrostatics and Pneumatics. With 175 Diagrams and Illustrations, and 522 Examination Questions and Answers. Crown 8vo., 2s. 6d.

    THEORETICAL MECHANICS—SOLIDS. With 163 Illustrations, 120 Worked Examples and over 500 Examples from Examination Papers, etc. Crown 8vo., 2s. 6d.

    THEORETICAL MECHANICS.—FLUIDS. With 122 Illustrations, numerous Worked Examples, and about 500 Examples from Examination Papers, etc. Crown 8vo., 2s. 6d.

THORNTON.—THEORETICAL MECHANICS—SOLIDS. Including Kinematics, Statics, and Kinetics. By ARTHUR THORNTON, M.A., F.R.A.S. With 200 Illustrations, 130 Worked Examples, and over 900 Examples from Examination Papers, etc. Crown 8vo., 4s. 6d.

MECHANICS, DYNAMICS, STATICS, HYDROSTATICS, ETC.—
Continued.

TWISDEN.—Works by the Rev. JOHN F. TWISDEN, M.A.

   PRACTICAL MECHANICS; an Elementary Introduction to their Study. With 855 Exercises, and 184 Figures and Diagrams. Crown 8vo., 10s. 6d.

   THEORETICAL MECHANICS. With 172 Examples, numerous Exercises, and 154 Diagrams. Crown 8vo., 8s. 6d.

WILLIAMSON.—INTRODUCTION TO THE MATHEMATICAL THEORY OF THE STRESS AND STRAIN OF ELASTIC SOLIDS. By BENJAMIN WILLIAMSON, D.Sc., F.R.S. Crown 8vo., 5s.

WILLIAMSON AND TARLETON.—AN ELEMENTARY TREATISE ON DYNAMICS. Containing Applications to Thermodynamics, with numerous Examples. By BENJAMIN WILLIAMSON, D.Sc., F.R.S., and FRANCIS A. TARLETON, LL.D. Crown 8vo., 10s. 6d.

WORTHINGTON.—DYNAMICS OF ROTATION: an Elementary Introduction to Rigid Dynamics. By A. M. WORTHINGTON, M.A., F.R.S. Crown 8vo., 3s. 6d.

OPTICS AND PHOTOGRAPHY.

ABNEY.—A TREATISE ON PHOTOGRAPHY. By Captain W. DE WIVELESLIE ABNEY, F.R.S., Director for Science in the Science and Art Department. With 115 Woodcuts. Fcp. 8vo., 3s. 6d.

GLAZEBROOK.—PHYSICAL OPTICS. By R. T. GLAZEBROOK, M.A., F.R.S., Fellow and Lecturer of Trinity College, Demonstrator of Physics at the Cavendish Laboratory, Cambridge. With 183 Woodcuts of Apparatus, etc. Fcp. 8vo., 6s.

WRIGHT.—OPTICAL PROJECTION: a Treatise on the Use of the Lantern in Exhibition and Scientific Demonstration. By LEWIS WRIGHT, Author of 'Light: a Course of Experimental Optics'. With 232 Illustrations. Crown 8vo., 6s.

SOUND, LIGHT, HEAT, AND THERMODYNAMICS.

ALEXANDER.—TREATISE ON THERMODYNAMICS. By PETER ALEXANDER, M.A. Crown 8vo., 5s.

CUMMING.—HEAT. For the Use of Schools and Students. By LINNÆUS CUMMING, M.A. With numerous Illustrations. Crown 8vo., 4s. 6d.

DAY.—NUMERICAL EXAMPLES IN HEAT. By R. E. DAY, M.A. Fcp. 8vo., 3s. 6d.

SOUND, LIGHT, HEAT, AND THERMODYNAMICS—*Continued*.

EMTAGE.—LIGHT. By W. T. A. EMTAGE, M.A. With 232 Illustrations. Crown 8vo., 6s.

HELMHOLTZ.—ON THE SENSATIONS OF TONE AS A PHYSIOLOGICAL BASIS FOR THE THEORY OF MUSIC. By HERMANN VON HELMHOLTZ. Royal 8vo., 28s.

MADAN.—AN ELEMENTARY TEXT-BOOK ON HEAT. For the Use of Schools. By H. G. MADAN, M.A., F.C.S., Fellow of Queen's College, Oxford; late Assistant Master at Eton College. Crown 8vo., 9s.

MAXWELL.—THEORY OF HEAT. By J. CLERK MAXWELL, M.A., F.R.SS., L. and E. With Corrections and Additions by Lord RAYLEIGH. With 38 Illustrations. Fcp. 8vo., 4s. 6d.

SMITH.—THE STUDY OF HEAT. By J. HAMBLIN SMITH, M.A., of Gonville and Caius College, Cambridge. Crown 8vo., 3s.

TYNDALL.—Works by JOHN TYNDALL, D.C.L., F.R.S. See p. 27.

WORMELL.—A CLASS-BOOK OF THERMODYNAMICS. By RICHARD WORMELL, B.Sc., M.A. Fcp. 8vo., 1s. 6d.

WRIGHT.—Works by MARK R. WRIGHT, Hon. Inter. B.Sc., London.

SOUND, LIGHT, AND HEAT. With 160 Diagrams and Illustrations. Crown 8vo., 2s. 6d.

ADVANCED HEAT. With 136 Diagrams and numerous Examples and Examination Papers. Crown 8vo., 4s. 6d.

STEAM, OIL, AND GAS ENGINES.

BALE.—A HAND-BOOK FOR STEAM USERS; being Rules for Engine Drivers and Boiler Attendants, with Notes on Steam Engine and Boiler Management and Steam Boiler Explosions. By M. POWIS BALE, M.I.M.E., A.M.I.C.E. Fcp. 8vo., 2s. 6d.

BOLTON.—MOTIVE POWERS AND THEIR PRACTICAL SELECTION. By REGINALD BOLTON, Associate Member of the Institution of Civil Engineers, etc. Crown 8vo., 6s. 6d. net.

CLERK.—THE GAS AND OIL ENGINE. By DUGALD CLERK, Associate Member of the Institution of Civil Engineers, Fellow of the Chemical Society, Member of the Royal Institution, Fellow of the Institute of Patent Agents. With 228 Illustrations and Diagrams. 8vo., 15s.

HOLMES.—THE STEAM ENGINE. By GEORGE C. V. HOLMES, Whitworth Scholar, Secretary of the Institution of Naval Architects. With 212 Woodcuts. Fcp. 8vo., 6s.

STEAM, OIL, AND GAS ENGINES—*Continued.*

NORRIS.—A PRACTICAL TREATISE ON THE 'OTTO' CYCLE GAS ENGINE. By WILLIAM NORRIS, M.I.Mech.E. With 207 Illustrations. 8vo., 10s. 6d.

RIPPER.—STEAM. By WILLIAM RIPPER, Member of the Institution of Mechanical Engineers; Professor of Mechanical Engineering in the Sheffield Technical School. With 142 Illustrations. Crown 8vo., 2s. 6d.

SENNETT.—THE MARINE STEAM ENGINE. A Treatise for the Use of Engineering Students and Officers of the Royal Navy. By RICHARD SENNETT, R.N., late Engineer-in-Chief of the Royal Navy. With 261 Illustrations. 8vo., 21s.

STROMEYER.—MARINE BOILER MANAGEMENT AND CONSTRUCTION. Being a Treatise on Boiler Troubles and Repairs, Corrosion, Fuels, and Heat, on the properties of Iron and Steel, on Boiler Mechanics, Workshop Practices, and Boiler Design. By C. E. STROMEYER, Graduate of the Royal Technical College at Aix-la-Chapelle, Member of the Institute of Naval Architects, etc. 8vo., 18s. net.

BUILDING CONSTRUCTION.

ADVANCED BUILDING CONSTRUCTION. By the Author of 'Rivingtons' Notes on Building Construction'. With 385 Illustrations. Crown 8vo., 4s. 6d.

BURRELL.—BUILDING CONSTRUCTION. By EDWARD J. BURRELL, Second Master of the People's Palace Technical School, London. With 303 Working Drawings. Crown 8vo., 2s. 6d.

SEDDON.—BUILDER'S WORK AND THE BUILDING TRADES. By Col. H. C. SEDDON, R.E., Superintending Engineer, H.M.'s Dockyard, Portsmouth; Examiner in Building Construction, Science and Art Department, South Kensington; with numerous Illustrations. Medium 8vo., 16s.

RIVINGTONS' COURSE OF BUILDING CONSTRUCTION.

NOTES ON BUILDING CONSTRUCTION. Arranged to meet the requirements of the syllabus of the Science and Art Department of the Committee of Council on Education, South Kensington. Medium 8vo.

    Part I. First Stage, or Elementary Course. With 552 Woodcuts, 10s. 6d.

    Part II. Commencement of Second Stage, or Advanced Course. With 479 Woodcuts, 10s. 6d.

    Part III. Materials. Advanced Course, and Course for Honours. With 188 Woodcuts, 21s.

    Part IV. Calculations for Building Structures. Course for Honours. With 597 Woodcuts, 15s.

ELECTRICITY AND MAGNETISM.

CUMMING.—ELECTRICITY TREATED EXPERIMEN-
TALLY. For the Use of Schools and Students. By LINNÆUS CUMMING, M.A. Assistant Master in Rugby School. With 242 Illustrations. Crown 8vo., 4*s.* 6*d.*

DAY.—EXERCISES IN ELECTRICAL AND MAGNETIC MEASUREMENTS, with Answers. By R. E. DAY. 12mo., 3*s.* 6*d.*

DU BOIS.—THE MAGNETIC CIRCUIT IN THEORY AND PRACTICE. By Dr. H. DU BOIS, Privatdocent in the University of Berlin. Translated by E. ATKINSON, Ph.D., formerly Professor of Experimental Science in the Staff College, Sandhurst. With 94 Illustrations. 8vo., 12*s.* net.

GORE.—THE ART OF ELECTRO-METALLURGY, including all known Processes of Electro-Deposition. By G. GORE, LL.D., F.R.S. With 56 Woodcuts. Fcp. 8vo., 6*s.*

JENKIN.—ELECTRICITY AND MAGNETISM. By FLEEMING JENKIN, F.R.S.S., L. and E., M.I.C.E. With 177 Illustrations. Fcp. 8vo., 3*s.* 6*d.*

JOUBERT.—ELEMENTARY TREATISE ON ELECTRICITY AND MAGNETISM. Founded on JOUBERT's 'Traité Élémentaire d'Electricité'. By G. C. FOSTER, F.R.S., Quain Professor of Physics in University College, London; and E. ATKINSON, Ph.D., formerly Professor of Experimental Science in the Staff College, Sandhurst. With 381 Illustrations. Crown 8vo., 7*s.* 6*d.*

LARDEN.—ELECTRICITY FOR PUBLIC SCHOOLS AND COLLEGES. By W. LARDEN, M.A. With 215 Illustrations, and a Series of Examination Papers, with Answers. Crown 8vo., 6*s.*

MERRIFIELD.—MAGNETISM AND DEVIATION OF THE COMPASS. For the Use of Students in Navigation and Science Schools. By JOHN MERRIFIELD, LL.D., F.R.A.S., 18mo., 2*s.* 6*d.*

POYSER.—Works by A. W. POYSER, M.A., Grammar School, Wisbeck.

   MAGNETISM AND ELECTRICITY. With 235 Illustrations. Crown 8vo., 2*s.* 6*d.*

   ADVANCED ELECTRICITY AND MAGNETISM. With 317 Illustrations. Crown 8vo., 4*s.* 6*d.*

SLINGO AND BROOKER.—Works by W. SLINGO and A. BROOKER.

   ELECTRICAL ENGINEERING FOR ELECTRIC LIGHT ARTISANS AND STUDENTS. With 346 Illustrations. Crown 8vo., 12*s.*

   PROBLEMS AND SOLUTIONS IN ELEMENTARY ELECTRICITY AND MAGNETISM. Embracing a Complete Set of Answers to the South Kensington Papers for the years 1885-1894, and a Series of Original Questions. With 67 Original Illustrations. Crown 8vo., 2*s.*

TYNDALL.—Works by JOHN TYNDALL, D.C.L., F.R.S. See p. 27.

TELEGRAPHY AND THE TELEPHONE.

BENNETT.—THE TELEPHONE SYSTEMS OF CONTINENTAL EUROPE. By A. R. BENNETT, Member of the Institute of Electrical Engineers; late General Manager in Scotland of the National Telephone Company, and General Manager and Electrician of the Mutual and New Telephone Companies. With 169 Illustrations. Crown 8vo., 15s.

CULLEY.—A HANDBOOK OF PRACTICAL TELEGRAPHY. By R. S. CULLEY, M.I.C.E., late Engineer-in-Chief of Telegraphs to the Post Office. With 135 Woodcuts and 17 Plates. 8vo., 16s.

PREECE AND SIVEWRIGHT.—TELEGRAPHY. By W. H. PREECE, C.B., F.R.S., V.P.Inst., C.E., etc., Engineer-in-Chief and Electrician Post Office Telegraphs; and Sir J. SIVEWRIGHT, K.C.M.G., General Manager, South African Telegraphs. With 258 Woodcuts. Fcp. 8vo., 6s.

ENGINEERING, STRENGTH OF MATERIALS, ETC.

ANDERSON.—THE STRENGTH OF MATERIALS AND STRUCTURES: the Strength of Materials as depending on their Quality and as ascertained by Testing Apparatus. By Sir J. ANDERSON, C.E., LL.D., F.R.S.E. With 66 Woodcuts. Fcp. 8vo., 3s. 6d.

BARRY.—RAILWAY APPLIANCES: a Description of Details of Railway Construction subsequent to the completion of the Earthworks and Structures. By JOHN WOLFE BARRY, C.B., M.I.C.E. With 218 Woodcuts. Fcp. 8vo., 4s. 6d.

SMITH.—GRAPHICS, or the Art of Calculation by Drawing Lines, applied especially to Mechanical Engineering. By ROBERT H. SMITH, Professor of Engineering, Mason College, Birmingham. Part I. With separate Atlas of 29 Plates containing 97 Diagrams. 8vo., 15s.

STONEY.—THE THEORY OF THE STRESSES ON GIRDERS AND SIMILAR STRUCTURES. With Practical Observations on the Strength and other Properties of Materials. By BINDON B. STONEY, LL.D., F.R.S., M.I.C.E. With 5 Plates and 143 Illustrations in the Text. Royal 8vo., 36s.

UNWIN.—Works by WILLIAM CAWTHORNE UNWIN, F.R.S., B.SC.

   THE TESTING OF MATERIALS OF CONSTRUCTION. Embracing the description of Testing Machinery and Apparatus Auxiliary to Mechanical Testing, and an Account of the most Important Researches on the Strength of Materials. With 141 Woodcuts and 5 Folding-out Plates. 8vo., 21s.

   ON THE DEVELOPMENT AND TRANSMISSION OF POWER FROM CENTRAL STATIONS: being the Howard Lectures delivered at the Society of Arts in 1893. With 81 Diagrams. 8vo., 10s. net.

WARREN.—ENGINEERING CONSTRUCTION IN IRON, STEEL, AND TIMBER. By WILLIAM HENRY WARREN, Challis Professor of Civil and Mechanical Engineering, University of Sydney. With 13 Folding Plates, and 375 Diagrams. Royal 8vo., 16s. net.

Scientific Works published by Longmans, Green, & Co. 13

MACHINE DRAWING AND DESIGN.

LOW AND BEVIS.—A MANUAL OF MACHINE DRAWING AND DESIGN. By DAVID ALLAN LOW (Whitworth Scholar), M.I.Mech.E., Head Master of the Technical School, People's Palace, London; and ALFRED WILLIAM BEVIS (Whitworth Scholar), M.I.Mech.E., Director of Manual Training to the Birmingham School Board. With over 700 Illustrations. 8vo., 7s. 6d.

LOW.—Works by DAVID ALLAN LOW, Head Master of the Technical School, People's Palace, London.

IMPROVED DRAWING SCALES. 4d. in case.

AN INTRODUCTION TO MACHINE DRAWING AND DESIGN. With 97 Illustrations and Diagrams. Crown 8vo., 2s.

UNWIN.—THE ELEMENTS OF MACHINE DESIGN. By W. CAWTHORNE UNWIN, F.R.S., Professor of Engineering at the Central Institute of the City and Guilds of London Institute.

Part I. General Principles, Fastenings, and Transmissive Machinery. With 304 Diagrams, etc. Crown 8vo., 6s.

Part II. Chiefly on Engine Details. With 174 Woodcuts. Crown 8vo., 4s. 6d.

LONGMANS' CIVIL ENGINEERING SERIES.

Edited by the Author of 'Notes on Building Construction'.

TIDAL RIVERS: their (1) Hydraulics, (2) Improvement, (3) Navigation. By W. H. WHEELER, M.Inst.C.E., author of 'The Drainage of Fens and Low Lands by Gravitation and Steam Power'. With 75 Illustrations. Medium 8vo., 16s. net.

NOTES ON DOCKS AND DOCK CONSTRUCTION. By C. COLSON, M.Inst.C.E., Assistant Director of Works, Admiralty. With 365 Illustrations. Medium 8vo., 21s. net.

PRINCIPLES AND PRACTICE OF HARBOUR CONSTRUCTION. By WILLIAM SHIELD, F.R.S.E., M.Inst.C.E., and Executive Engineer, National Harbour of Refuge, Peterhead, N.B. With 97 Illustrations. Medium 8vo., 15s. net.

CALCULATIONS FOR ENGINEERING STRUCTURES. By T. CLAXTON FIDLER, M.I.C.E., Professor of Engineering in the University of Dundee; Author of 'A Practical Treatise on Bridge Construction'.
[In preparation.

A COURSE OF CIVIL ENGINEERING. By L. F. VERNON-HARCOURT, M.Inst.C.E., Professor of Civil Engineering at University College, London.
[In preparation.

RAILWAY CONSTRUCTION. By W. H. MILLS, M.I.C.E., Engineer-in-Chief, Great Northern Railway, Ireland.
[In preparation.

WORKSHOP APPLIANCES, ETC.

NORTHCOTT.—LATHES AND TURNING, Simple, Mechanical and Ornamental. By W. H. NORTHCOTT. With 338 Illustrations. 8vo., 18s.

SHELLEY.—WORKSHOP APPLIANCES, including Descriptions of some of the Gauging and Measuring Instruments, Hand-cutting Tools, Lathes, Drilling, Plaining, and other Machine Tools used by Engineers. By C. P. B. SHELLEY, M.I.C.E. With an additional Chapter on Milling by R. R. LISTER. With 323 Woodcuts. Fcp. 8vo., 5s.

MINERALOGY, METALLURGY, ETC.

BAUERMAN.—Works by HILARY BAUERMAN, F.G.S.
  SYSTEMATIC MINERALOGY. With 373 Woodcuts and Diagrams. Fcp. 8vo., 6s.
  DESCRIPTIVE MINERALOGY. With 236 Woodcuts and Diagrams. Fcp. 8vo., 6s.

BLOXAM AND HUNTINGTON.—METALS: their Properties and Treatment. By C. L. BLOXAM and A. K. HUNTINGTON, Professors in King's College, London. With 130 Woodcuts. Fcp. 8vo., 5s.

GORE.—THE ART OF ELECTRO-METALLURGY, including all known Processes of Electro-Deposition. By G. GORE, LL.D., F.R.S. With 56 Woodcuts. Fcp. 8vo., 6s.

MITCHELL.—A MANUAL OF PRACTICAL ASSAYING. By JOHN MITCHELL, F.C.S. Revised, with the Recent Discoveries incorporated. By W. CROOKES, F.R.S. With 201 Illustrations. 8vo., 31s. 6d.

RHEAD.—METALLURGY. An Elementary Text-Book. By E. C. RHEAD, Lecturer on Metallurgy at the Municipal Technical School, Manchester. With 94 Illustrations. Crown 8vo., 3s. 6d.

RUTLEY.—THE STUDY OF ROCKS: an Elementary Text-book of Petrology. By F. RUTLEY, F.G.S. With 6 Plates and 88 Woodcuts. Fcp. 8vo., 4s. 6d.

ASTRONOMY, NAVIGATION, ETC.

ABBOTT.—ELEMENTARY THEORY OF THE TIDES: the Fundamental Theorems Demonstrated without Mathematics and the Influence on the Length of the Day Discussed. By T. K. ABBOTT, B.D., Fellow and Tutor, Trinity College, Dublin. Crown 8vo., 2s.

BALL.—Works by Sir ROBERT S. BALL, LL.D., F.R.S.
  ELEMENTS OF ASTRONOMY. With 130 Figures and Diagrams. Fcp. 8vo., 6s. 6d.
  A CLASS-BOOK OF ASTRONOMY. With 41 Diagrams. Fcp. 8vo., 1s. 6d.

Scientific Works published by Longmans, Green, & Co. 15

ASTRONOMY, NAVIGATION, ETC.—*Continued.*

BRINKLEY.—ASTRONOMY. By F. BRINKLEY, formerly Astronomer Royal for Ireland. Re-edited and Revised by J. W. STUBBS, D.D. and F. BRÜNNOW, Ph.D. With 49 Diagrams. Crown 8vo., 6*s*.

CLERKE.—THE SYSTEM OF THE STARS. By AGNES M. CLERKE. With 6 Plates, and numerous Illustrations. 8vo., 21*s*.

GOODWIN.—AZIMUTH TABLES FOR THE HIGHER DECLINATIONS. (Limits of Declination 24° to 30°, both inclusive.) Between the Parallels of Latitude 0° and 60°. With Examples of the Use of the Tables in English and French. By H. B. GOODWIN, Naval Instructor, Royal Navy. Royal 8vo., 7*s*. 6*d*.

HERSCHEL.—OUTLINES OF ASTRONOMY.—By Sir JOHN F. W. HERSCHEL, Bart., K.H., etc. With 9 Plates, and numerous Diagrams. 8vo., 12*s*.

LOWELL.—MARS. By PERCIVAL LOWELL, Fellow American Academy, Member Royal Asiatic Society, Great Britain and Ireland, etc. With 24 Plates. 8vo., 12*s*. 6*d*.

MARTIN.—NAVIGATION AND NAUTICAL ASTRONOMY. Compiled by Staff Commander W. R. MARTIN, R.N. Royal 8vo., 18*s*.

MERRIFIELD.—A TREATISE ON NAVIGATION. For the Use of Students. By J. MERRIFIELD, LL.D., F.R.A.S., F.M.S. With Charts and Diagrams. Crown 8vo., 5*s*.

PARKER.—ELEMENTS OF ASTRONOMY. With Numerous Examples and Examination Papers. By GEORGE W. PARKER, M.A., of Trinity College, Dublin. With 84 Diagrams. 8vo., 5*s*. net.

WEBB.—CELESTIAL OBJECTS FOR COMMON TELESCOPES. By the Rev. T. W. WEBB, M.A., F.R.A.S. Fifth Edition, Revised and greatly Enlarged by the Rev. T. E. ESPIN, M.A., F.R.A.S. (Two Volumes.) Vol. I., with Portrait and a Reminiscence of the Author, 2 Plates, and numerous Illustrations. Crown 8vo., 6*s*. Vol. II., with numerous Illustrations. Crown 8vo., 6*s*. 6*d*.

WORKS BY RICHARD A. PROCTOR.

OLD AND NEW ASTRONOMY. With 21 Plates and 472 Illustrations in the Text. 4to., 21*s*.

MYTHS AND MARVELS OF ASTRONOMY. Crown 8vo., 3*s*. 6*d*.

THE MOON: Her Motions, Aspect, Scenery, and Physical Condition. With many Plates and Charts, Wood Engravings, and 2 Lunar Photographs. Crown 8vo., 5*s*.

THE UNIVERSE OF STARS: Researches into, and New Views respecting the Constitution of the Heavens. With 22 Charts (4 Coloured), and 22 Diagrams. 8vo., 10*s*. 6*d*.

[OVER.

WORKS BY RICHARD A. PROCTOR—*Continued*.

OTHER WORLDS THAN OURS: the Plurality of Worlds Studied Under the Light of Recent Scientific Researches. With 14 Illustrations; Map, Charts, etc. Crown 8vo., 3s. 6d.

THE ORBS AROUND US; Essays on the Moon and Planets, Meteors and Comets, the Sun and Coloured Pairs of Suns. Crown 8vo., 3s. 6d.

LIGHT SCIENCE FOR LEISURE HOURS: Familiar Essays on Scientific Subjects. Natural Phenomena, etc. 3 vols., crown 8vo., 5s. each

THE EXPANSE OF HEAVEN: Essays on the Wonders of the Firmament. Crown 8vo., 3s. 6d.

OTHER SUNS THAN OURS: a Series of Essays on Suns—Old, Young, and Dead. With other Science Gleanings. Two Essays on Whist, and Correspondence with Sir John Herschel. With 9 Star-Maps and Diagrams. Crown 8vo., 3s. 6d.

HALF-HOURS WITH THE TELESCOPE: a Popular Guide to the Use of the Telescope as a means of Amusement and Instruction. With 7 Plates. Fcp. 8vo., 2s. 6d.

NEW STAR ATLAS FOR THE LIBRARY, the School, and the Observatory, in Twelve Circular Maps (with Two Index-Plates). With an Introduction on the Study of the Stars. Illustrated by 9 Diagrams. Crown 8vo., 5s.

THE SOUTHERN SKIES: a Plain and Easy Guide to the Constellations of the Southern Hemisphere. Showing in 12 Maps the position of the principal Star-Groups night after night throughout the year. With an Introduction and a separate Explanation of each Map. True for every Year. 4to., 5s.

HALF-HOURS WITH THE STARS: a Plain and Easy Guide to the Knowledge of the Constellations. Showing in 12 Maps the position of the principal Star-Groups night after night throughout the year. With Introduction and a separate Explanation of each Map. True for every Year. 4to., 3s. 6d.

LARGER STAR ATLAS FOR OBSERVERS AND STUDENTS. In Twelve Circular Maps, showing 6000 Stars, 1500 Double Stars, Nebulæ, etc. With 2 Index-Plates. Folio, 15s.

THE STARS IN THEIR SEASONS: an Easy Guide to a Knowledge of the Star-Groups. In 12 Large Maps. Imperial 8vo., 5s.

ROUGH WAYS MADE SMOOTH. Familiar Essays on Scientific Subjects. Crown 8vo., 3s. 6d.

PLEASANT WAYS IN SCIENCE. Crown 8vo., 3s. 6d.

NATURE STUDIES. By R. A. PROCTOR, GRANT ALLEN, A. WILSON, T. FOSTER, and E. CLODD. Crown 8vo., 3s. 6d.

LEISURE READINGS. By R. A. PROCTOR, E. CLODD, A. WILSON, T. FOSTER, and A. C. RANYARD. Crown 8vo., 3s. 6d.

Scientific Works published by Longmans, Green, & Co. 17

MANUFACTURES, TECHNOLOGY, ETC.

BELL.—JACQUARD WEAVING AND DESIGNING. By F. T. BELL, Medallist in Honours and Certificated Teacher in 'Linen Manufacturing' and in 'Weaving and Pattern Designing,' City and Guilds of London Institute. With 199 Diagrams. 8vo., 12s. net.

CROSS AND BEVAN.—CELLULOSE: an Outline of the Chemistry of the Structural Elements of Plants. With Reference to their Natural History and Industrial Uses. By CROSS and BEVAN (C. F. Cross, E. J. Bevan, and C. Beadle). With 14 Plates. Crown 8vo., 12s. net.

TAYLOR.—COTTON WEAVING AND DESIGNING. By JOHN T. TAYLOR. With 373 Diagrams. Crown 8vo., 7s. 6d. net.

LUPTON.—MINING. An Elementary Treatise on the Getting of Minerals. By ARNOLD LUPTON, M.I.C.E., F.G.S., etc. With 596 Diagrams and Illustrations. Crown 8vo., 9s. net.

WATTS.—AN INTRODUCTORY MANUAL FOR SUGAR GROWERS. By FRANCIS WATTS, F.C.S., F.I.C. With 20 Illustrations. Crown 8vo., 6s.

PHYSIOGRAPHY AND GEOLOGY.

BIRD.—Works by CHARLES BIRD, B.A.

ELEMENTARY GEOLOGY. With Geological Map of the British Isles, and 247 Illustrations. Crown 8vo., 2s. 6d.

ADVANCED GEOLOGY. A Manual for Students in Advanced Classes and for General Readers. With over 300 Illustrations, a Geological Map of the British Isles (coloured), and a set of Questions for Examination. Crown 8vo., 7s. 6d.

GREEN.—PHYSICAL GEOLOGY FOR STUDENTS AND GENERAL READERS. With Illustrations. By A. H. GREEN, M.A., F.G.S. 8vo., 21s.

THORNTON.—Works by J. THORNTON, M.A.

ELEMENTARY PHYSIOGRAPHY: an Introduction to the Study of Nature. With 12 Maps and 247 Illustrations. With Appendix on Astronomical Instruments and Measurements. Crown 8vo., 2s. 6d.

ADVANCED PHYSIOGRAPHY. With 6 Maps and 180 Illustrations. Crown 8vo., 4s. 6d.

HEALTH AND HYGIENE.

BRAY.—PHYSIOLOGY AND THE LAWS OF HEALTH, in Easy Lessons for Schools. By Mrs. CHARLES BRAY. Fcp. 8vo., 1s.

BRODRIBB.—MANUAL OF HEALTH AND TEMPERANCE. By T. BRODRIBB, M.A. With Extracts from Gough's 'Temperance Orations'. Revised and Edited by the Rev. W. RUTHVEN PYM, M.A., Member of the Sheffield School Board. Crown 8vo., 1s. 6d.

BUCKTON. — HEALTH IN THE HOUSE; Twenty-five Lectures on Elementary Physiology. By Mrs. C. M. BUCKTON. With 41 Woodcuts and Diagrams. Crown 8vo., 2s.

HEALTH AND HYGIENE—*Continued.*

CORFIELD.—THE LAWS OF HEALTH. By W. H. Corfield, M.A., M.D. Fcp. 8vo., 1s. 6d.

FRANKLAND.—MICRO-ORGANISMS IN WATER, THEIR SIGNIFICANCE, IDENTIFICATION, AND REMOVAL. Together with an Account of the Bacteriological Methods Involved in their Investigation. Specially Designed for the Use of those connected with the Sanitary Aspects of Water Supply. By Professor Percy Frankland, Ph.D., B.Sc. (Lond.), F.R.S., Fellow of the Chemical Society; and Mrs. Percy Frankland. With 2 Plates and numerous Diagrams. 8vo., 16s. net.

NOTTER AND FIRTH.—HYGIENE. By J. L. Notter, M.A., M.D., and R. H. Firth, F.R.C.S. With 95 Illustrations. Crown 8vo., 3s. 6d.

POORE.—ESSAYS ON RURAL HYGIENE. By George Vivian Poore, M.D. Crown 8vo., 6s. 6d.

WILSON.—A MANUAL OF HEALTH-SCIENCE: adapted for use in Schools and Colleges, and suited to the requirements of Students preparing for the Examinations in Hygiene of the Science and Art Department, etc. By Andrew Wilson, F.R.S.E., F.L.S., etc. With 74 Illustrations. Crown 8vo., 2s. 6d.

NATURAL HISTORY, EVOLUTION, ETC.

CLODD.—Works by EDWARD CLODD.

THE STORY OF CREATION: A Plain Account of Evolution. With 77 Illustrations. Crown 8vo., 3s. 6d.

A PRIMER OF EVOLUTION: being a Popular Abridged Edition of 'The Story of Creation'. With Illustrations. Fcp. 8vo., 1s. 6d.

FURNEAUX.—Works by WILLIAM S. FURNEAUX.

THE OUTDOOR WORLD; or, The Young Collector's Handbook. With 18 Plates, 16 of which are coloured, and 549 Illustrations in the Text. Crown 8vo., 7s. 6d.

BUTTERFLIES AND MOTHS (British). With 12 Coloured Plates and 241 Illustrations in the Text. 12s. 6d.

LIFE IN PONDS AND STREAMS. With 8 Coloured Plates and 311 Illustrations in the Text. Crown 8vo., 12s. 6d.

HUDSON.—BRITISH BIRDS. By W. H. Hudson, C.M.Z.S. With 8 Coloured Plates from Original Drawings by A. Thorburn, and 8 Plates and 100 Figures in black and white from Original Drawings by C. E. Lodge, and 3 Illustrations from Photographs by R. B. Lodge. Crown 8vo., 12s. 6d.

ROMANES.—Works by GEORGE JOHN ROMANES, LL.D., F.R.S.

DARWIN, AND AFTER DARWIN: an Exposition of the Darwinian Theory, and a Discussion on Post-Darwinian Questions.

    Part I. THE DARWINIAN THEORY. With Portrait of Darwin and 125 Illustrations. Crown 8vo., 10s. 6d.

    Part II. POST-DARWINIAN QUESTIONS: Heredity and Utility. With Portrait of the Author and 5 Illustrations. Crown 8vo., 10s. 6d.

AN EXAMINATION OF WEISMANNISM. Crown 8vo., 6s.

MEDICINE AND SURGERY.

ASHBY AND **WRIGHT.**—THE DISEASES OF CHILDREN, MEDICAL AND SURGICAL. By HENRY ASHBY, M.D., Lond., F.R.C.P., and G. A. WRIGHT, B.A., M.B., Oxon., F.R.C.S., Eng. With 192 Illustrations. 8vo., 25s.

BENNETT.—Works by WILLIAM H. BENNETT, F.R.C.S., Surgeon to St. George's Hospital.

CLINICAL LECTURES ON VARICOSE VEINS OF THE LOWER EXTREMITIES. With 3 Plates. 8vo., 6s.

ON VARICOCELE; A PRACTICAL TREATISE. With 4 Tables and a Diagram. 8vo., 5s.

CLINICAL LECTURES ON ABDOMINAL HERNIA: chiefly in relation to Treatment, including the Radical Cure. With 12 Diagrams in the Text. 8vo., 8s. 6d.

CLARKE. — POST-MORTEM EXAMINATIONS IN MEDICO-LEGAL AND ORDINARY CASES. With Special Chapters on the Legal Aspects of Post-Mortems, and on Certificates of Death. By J. JACKSON CLARKE, M.B. (Lond.), F.R.C.S. Fcp. 8vo., 2s. 6d.

COATS.—A MANUAL OF PATHOLOGY. By JOSEPH COATS, M.D., Professor of Pathology in the University of Glasgow. Third Edition. Revised throughout. With 507 Illustrations. 8vo., 31s. 6d.

COOKE.—Works by THOMAS COOK, F.R.C.S. Eng., B.A., B.Sc., M.D., Paris.

TABLETS OF ANATOMY. Being a Synopsis of Demonstrations given in the Westminster Hospital Medical School in the years 1871-75. Tenth Thousand, being a selection of the Tablets believed to be most useful to Students generally. Post 4to., 10s. 6d.

APHORISMS IN APPLIED ANATOMY AND OPERATIVE SURGERY. Including 100 Typical *vivâ voce* Questions on Surface Marking, etc. Crown 8vo., 3s. 6d.

DISSECTION GUIDES. Aiming at Extending and Facilitating such Practical work in Anatomy as will be specially useful in connection with an ordinary Hospital Curriculum. 8vo., 10s. 6d.

DICKINSON.—Works by W. HOWSHIP DICKINSON, M.D. Cantab., F.R.C.P.

ON RENAL AND URINARY AFFECTIONS. Complete in Three Parts, 8vo., with 12 Plates and 122 Woodcuts. Price £3 4s. 6d. cloth.

*** The Parts can also be had separately, each complete in itself, as follows:—

Part I. Diabetes, 10s. 6d. sewed, 12s. cloth.

Part II. Albuminuria, £1 sewed, £1 1s. cloth.

Part III. Miscellaneous Affections of the Kidneys and Urine, £1 10s. sewed, £1 11s. 6d. cloth.

MEDICINE AND SURGERY—*Continued*.

DICKINSON.—Works by W. HOWSHIP DICKINSON, M.D. Cantab., F.R.C.P.—*continued.*

THE TONGUE AS AN INDICATION OF DISEASE; being the Lumleian Lectures delivered at the Royal College of Physicians in March, 1888. 8vo., 7s. 6d.

THE HARVEIAN ORATION ON HARVEY IN ANCIENT AND MODERN MEDICINE. Crown 8vo., 2s. 6d.

OCCASIONAL PAPERS ON MEDICAL SUBJECTS, 1855-1896. 8vo., 12s.

ERICHSEN.—**THE SCIENCE AND ART OF SURGERY;** a Treatise on Surgical Injuries, Diseases, and Operations. Tenth Edition. Revised by the late MARCUS BECK, M.S., and M.B. (Lond.), F.R.C.S., and by RAYMOND JOHNSON, M.B. and B.S. (Lond.), F.R.C.S. Illustrated by nearly 1000 Engravings on Wood. 2 vols., royal 8vo., £8s.

GARROD.—Works by SIR ALFRED BARING GARROD, M.D., F.R.S., etc.

A TREATISE ON GOUT AND RHEUMATIC GOUT (RHEUMATOID ARTHRITIS). With 6 Plates, comprising 21 Figures (14 Coloured), and 27 Illustrations engraved on Wood. 8vo., 21s.

THE ESSENTIALS OF MATERIA MEDICA AND THERAPEUTICS. Revised and Edited, under the supervision of the Author, by NESTOR TIRARD, M.D., Lond., F.R.C.P., Professor of Materia Medica and Therapeutics in King's College, London, etc. Crown 8vo., 12s. 6d.

GRAY.—**ANATOMY, DESCRIPTIVE AND SURGICAL.** By HENRY GRAY, F.R.S., late Lecturer on Anatomy at St. George's Hospital. The Thirteenth Edition, re-edited by T. PICKERING PICK, Surgeon to St. George's Hospital; Member of the Court of Examiners, Royal College of Surgeons of England. With 636 large Woodcut Illustrations, a large proportion of which are coloured, the Arteries being coloured red, the Veins blue, and the Nerves yellow. The attachments of the muscles to the bones, in the section on Osteology, are also shown in coloured outline. Royal 8vo., 36s.

HALLIBURTON.—Works by W. D. HALLIBURTON, M.D., F.R.S., M.R.C.P.

A TEXT-BOOK OF CHEMICAL PHYSIOLOGY AND PATHOLOGY. With 104 Illustrations. 8vo., 28s.

ESSENTIALS OF CHEMICAL PHYSIOLOGY. Second Edition. 8vo., 5s.

LANG.—**THE METHODICAL EXAMINATION OF THE EYE.** Being Part I. of a Guide to the Practice of Ophthalmology for Students and Practitioners. By WILLIAM LANG, F.R.C.S., Eng., Surgeon to the Royal London Ophthalmic Hospital, Moorfields, etc. With 15 Illustrations. Crown 8vo., 3s. 6d.

Scientific Works published by Longmans, Green, & Co. 21

MEDICINE AND SURGERY—*Continued.*

LIVEING.—Works by ROBERT LIVEING, M.A. and M.D. Cantab., F.R.C.P., Lond., etc., Physician to the Department for Diseases of the Skin at the Middlesex Hospital, etc.

HANDBOOK ON DISEASES OF THE SKIN. With especial reference to Diagnosis and Treatment. Fcap. 8vo., 5s.

ELEPHANTIASIS GRÆCORUM, OR TRUE LEPROSY; Being the Goulstonian Lectures for 1873. Cr. 8vo., 4s. 6d.

LONGMORE.—Works by Surgeon-General Sir T. LONGMORE (Retired), C.B., F.R.C.S.

THE ILLUSTRATED OPTICAL MANUAL; OR, HANDBOOK OF INSTRUCTIONS FOR THE GUIDANCE OF SURGEONS IN TESTING QUALITY AND RANGE OF VISION, AND IN DISTINGUISHING AND DEALING WITH OPTICAL DEFECTS IN GENERAL. Illustrated by 74 Drawings and Diagrams by Inspector-General Dr. MACDONALD, R.N., F.R.S., C.B. 8vo., 14s.

GUNSHOT INJURIES. Their History, Characteristic Features, Complications, and General Treatment; with Statistics concerning them as they have been met with in Warfare. With 78 Illustrations. 8vo., 31s. 6d.

LUFF.—TEXT-BOOK OF FORENSIC MEDICINE AND TOXICOLOGY. By ARTHUR P. LUFF, M.D., B.Sc. (Lond.), Physician in Charge of Out-Patients and Lecturer on Medical Jurisprudence and Toxicology in St. Mary's Hospital; Examiner in Forensic Medicine in the University of London; External Examiner in Forensic Medicine in the Victoria University; Official Analyst to the Home Office. With numerous Illustrations. 2 vols., crown 8vo., 24s.

NEWMAN.—ON THE DISEASES OF THE KIDNEY AMENABLE TO SURGICAL TREATMENT. Lectures to Practitioners. By DAVID NEWMAN, M.D., Surgeon to the Western Infirmary Out-Door Department; Pathologist and Lecturer on Pathology at the Glasgow Royal Infirmary; Examiner in Pathology in the University of Glasgow; Vice-President Glasgow Pathological and Clinical Society. 8vo., 16s.

OWEN.—A MANUAL OF ANATOMY FOR SENIOR STUDENTS. By EDMUND OWEN, M.B., F.R.S.C., Senior Surgeon to the Hospital for Sick Children, Great Ormond Street, Surgeon to St. Mary's Hospital, London, and co-Lecturer on Surgery, late Lecturer on Anatomy in its Medical School. With 210 Illustrations. Crown 8vo., 12s. 6d.

POOLE.—COOKERY FOR THE DIABETIC. By W. H and Mrs. POOLE. With Preface by Dr. PAVY. Fcap. 8vo., 2s. 6d.

MEDICINE AND SURGERY—*Continued.*

QUAIN.—A DICTIONARY OF MEDICINE; Including General Pathology, General Therapeutics, Hygiene, and the Diseases of Women and Children. By Various Writers. Edited by RICHARD QUAIN, Bart., M.D. Lond., LL.D. Edin. (Hon.) F.R.S., Physician Extraordinary to H.M. the Queen, President of the General Medical Council, Member of the Senate of the University of London, etc. Assisted by FREDERICK THOMAS ROBERTS, M.D. Lond., B.Sc., Fellow of the Royal College of Physicians, Fellow of University College, Professor of Materia Medica and Therapeutics, University College, &c.; and J. MITCHELL BRUCE, M.A. Abdn., M.D. Lond., Fellow of the Royal College of Physicians of London, Physician and Lecturer on the Principles and Practice of Medicine, Charing Cross Hospital, &c. New Edition, Revised throughout and Enlarged. In 2 Vols., medium 8vo., cloth, red edges, price 40s. net.

QUAIN.—QUAIN'S (JONES) ELEMENTS OF ANATOMY. The Tenth Edition. Edited by EDWARD ALBERT SCHÄFER, F.R.S., Professor of Physiology and Histology in University College, London; and GEORGE DANCER THANE, Professor of Anatomy in University College, London. In 3 Vols.

*** The several parts of this work form COMPLETE TEXT-BOOKS OF THEIR RESPECTIVE SUBJECTS. They can be obtained separately as follows:—

VOL. I, PART I. EMBRYOLOGY. By E. A. SCHÄFER, F.R.S. With 200 Illustrations. Royal 8vo., 9s.

VOL. I., PART II. GENERAL ANATOMY OR HISTOLOGY. By E. A. SCHÄFER, F.R.S. With 291 Illustrations. Royal 8vo., 12s. 6d.

VOL. II., PART I. OSTEOLOGY. By G. D. THANE. With 168 Illustrations. Royal 8vo., 9s.

VOL. II., PART II. ARTHROLOGY —MYOLOGY — ANGEIOLOGY. By G. D. THANE. With 255 Illustrations. Royal 8vo., 18s.

VOL. III., PART I. THE SPINAL CORD AND BRAIN. By E. A. SCHÄFER, F.R.S. With 139 Illustrations. Royal 8vo., 12s. 6d.

VOL. III. PART II. THE NERVES. By G. D. THANE. With 102 Illustrations. Royal 8vo., 9s.

VOL. III., PART III. THE ORGANS OF THE SENSES. By E. A. SCHÄFER, F.R.S. With 178 Illustrations. Royal 8vo., 9s.

VOL. III., PART IV. SPLANCHNOLOGY. By E. A. SCHÄFER, F.R.S., and JOHNSON SYMINGTON. M.D. With 337 Illustrations. Royal 8vo., 16s.

APPENDIX. SUPERFICIAL AND SURGICAL ANATOMY. By Professor G. D. THANE and Professor R. J. GODLEE, M.S. With 29 Illustrations. Royal 8vo., 6s. 6d.

SCHÄFER.—THE ESSENTIALS OF HISTOLOGY. Descriptive and Practical. For the Use of Students. By E. A. SCHÄFER, F.R.S., Jodrell Professor of Physiology in University College, London; Editor of the Histological Portion of Quain's 'Anatomy'. Illustrated by more than 300 Figures, many of which are new. 8vo., 7s. 6d. (Interleaved, 10s.)

SCHENK.—MANUAL OF BACTERIOLOGY. For Practitioners and Students. With especial reference to Practical Methods. By Dr. S. L. SCHENK, Professor (Extraordinary) in the University of Vienna. Translated from the German, with an Appendix, by W. R. DAWSON, B.A., M.D., Univ. Dub.; late University Travelling Prizeman in Medicine. With 100 Illustrations, some of which are coloured. 8vo., 10s. net.

Scientific Works published by Longmans, Green, & Co. 23

MEDICINE AND SURGERY—*Continued.*

SMALE AND COLYER. DISEASES AND INJURIES OF
THE TEETH, including Pathology and Treatment: a Manual of Practical Dentistry for Students and Practitioners. By MORTON SMALE, M.R.C.S., L.S.A., L.D.S., Dental Surgeon to St. Mary's Hospital, Dean of the School, Dental Hospital of London, etc.; and J. F. COLYER, L.R.C.P., M.R.C.S., L.D.S., Assistant Dental Surgeon to Charing Cross Hospital, and Assistant Dental Surgeon to the Dental Hospital of London. With 334 Illustrations. Large Crown 8vo., 15s.

SMITH (H. F.). THE HANDBOOK FOR MIDWIVES By HENRY FLY SMITH, B.A., M.B. Oxon., M.R.C.S. With 41 Woodcuts. Crown 8vo., price 5s.

STEEL.—Works by JOHN HENRY STEEL, F.R.C.V.S., F.Z.S., A.V.D.

A TREATISE ON THE DISEASES OF THE DOG; being a Manual of Canine Pathology. Especially adapted for the use of Veterinary Practitioners and Students. 88 Illustrations. 8vo., 10s 6d.

A TREATISE ON THE DISEASES OF THE OX; being a Manual of Bovine Pathology. Especially adapted for the use of Veterinary Practitioners and Students. 2 Plates and 117 Woodcuts. 8vo., 15s.

A TREATISE ON THE DISEASES OF THE SHEEP; being a Manual of Ovine Pathology for the use of Veterinary Practitioners and Students. With Coloured Plate and 99 Woodcuts. 8vo., 12s.

OUTLINES OF EQUINE ANATOMY; a Manual for the use of Veterinary Students in the Dissecting Room. Crown 8vo., 7s. 6d.

'STONEHENGE.'—THE DOG IN HEALTH AND DISEASE. By 'STONEHENGE'. With 84 Wood Engravings. Square Crown 8vo., 7s. 6d.

WALLER.—AN INTRODUCTION TO HUMAN PHYSIOLOGY. By AUGUSTUS D. WALLER, M.D., Lecturer on Physiology at St. Mary's Hospital Medical School, London; late External Examiner at the Victorian University. Second Edition, Revised. With 305 Illustrations. 8vo., 18s.

WEICHSELBAUM.—THE ELEMENTS OF PATHOLOGICAL HISTOLOGY, With Special Reference to Practical Methods. By Dr. ANTON WEICHSELBAUM, Professor of Pathology in the University of Vienna. Translated by W. R. DAWSON, M.D. (Dub.), Demonstrator of Pathology in the Royal College of Surgeons, Ireland, late Medical Travelling Prizeman of Dublin University, &c. With 221 Figures, partly in Colours, a Cromo-lithographic Plate, and 7 Photographic Plates. Royal 8vo., 21s. net.

MEDICINE AND SURGERY—*Continued.*

WILKS AND MOXON.—LECTURES ON PATHOLOGICAL ANATOMY. By SAMUEL WILKS, M.D., F.R.S., Consulting Physician to, and formerly Lecturer on Medicine and Pathology at Guy's Hospital, and the late WALTER MOXON, M.D., F.R.C.P., Physician to, and some time Lecturer on Pathology at Guy's Hospital. Third Edition, thoroughly Revised. By SAMUEL WILKS, M.D., LL.D., F.R.S. 8vo., 18s.

YOUATT.—Works by WILLIAM YOUATT.

 THE HORSE. Revised and Enlarged by W. WATSON, M.R.C.V.S. With 52 Woodcuts. 8vo., 7s. 6d.

 THE DOG. Revised and Enlarged. With 33 Woodcuts. 8vo., 6s.

PHYSIOLOGY, BIOLOGY, ETC.

ASHBY.—NOTES ON PHYSIOLOGY, for the Use of Students Preparing for Examination. By HENRY ASHBY, M.D. With 141 Illustrations. Fcp. 8vo., 5s.

BARNETT.—THE MAKING OF THE BODY: a Children's Book on Anatomy and Physiology, for School and Home Use. By Mrs. S. A. BARNETT. With 113 Illustrations. Crown 8vo., 1s. 9d.

BIDGOOD.—A COURSE OF PRACTICAL ELEMENTARY BIOLOGY. By JOHN BIDGOOD, B.Sc., F.L.S. With 226 Illustrations. Crown 8vo., 4s. 6d.

BRAY.—PHYSIOLOGY AND THE LAWS OF HEALTH, in Easy Lessons for Schools. By Mrs. CHARLES BRAY. Fcp. 8vo., 1s.

FRANKLAND.—MICRO-ORGANISMS IN WATER. Together with an Account of the Bacteriological Methods involved in their Investigation. Specially designed for the use of those connected with the Sanitary Aspects of Water-Supply. By PERCY FRANKLAND, Ph.D., B.Sc. (Lond.), F.R.S., and Mrs. PERCY FRANKLAND. With 2 Plates and Numerous Diagrams. 8vo., 16s. net.

FURNEAUX.—HUMAN PHYSIOLOGY. By W. FURNEAUX, F.R.G.S. With 218 Illustrations. Crown 8vo., 2s. 6d.

HUDSON AND GOSSE.—THE ROTIFERA, or 'WHEEL-ANIMALCULES'. By C. T. HUDSON, LL.D., and P. H. GOSSE, F.R.S. With 30 Coloured and 4 Uncoloured Plates. In 6 Parts. 4to., 10s. 6d. each; Supplement 12s. 6d. Complete in 2 vols., with Supplement, 4to., £4 4s.

Scientific Works published by Longmans, Green, & Co. 25

PHYSIOLOGY, BIOLOGY, ETC.—*Continued.*

MACALISTER.—Works by ALEXANDER MACALISTER, M.D.

   ZOOLOGY AND MORPHOLOGY OF VERTEBRATA.
      8vo., 10s. 6d.

   ZOOLOGY OF THE INVERTEBRATE ANIMALS. With
      59 Diagrams. Fcp. 8vo., 1s. 6d.

   ZOOLOGY OF THE VERTEBRATE ANIMALS. With 77
      Diagrams. Fcp. 8vo., 1s. 6d.

MORGAN.—ANIMAL BIOLOGY: an Elementary Text-Book. By C. LLOYD MORGAN. With numerous Illustrations. Crown 8vo., 8s. 6d.

SCHENK.—MANUAL OF BACTERIOLOGY, for Practitioners and Students, with Especial Reference to Practical Methods. By Dr. S. L. SCHENK. With 100 Illustrations, some of which are Coloured. 8vo., 10s. net.

THORNTON.—HUMAN PHYSIOLOGY. By JOHN THORNTON, M.A. With 267 Illustrations, some Coloured. Crown 8vo., 6s.

BOTANY.

AITKEN.—ELEMENTARY TEXT-BOOK OF BOTANY. For the use of Schools. By EDITH AITKEN, late Scholar of Girton College. With over 400 Diagrams. Crown 8vo., 4s. 6d.

BENNETT AND MURRAY.—HANDBOOK OF CRYPTO-GAMIC BOTANY. By ALFRED W. BENNETT, M.A., B.Sc., F.L.S., Lecturer on Botany at St. Thomas's Hospital; and GEORGE MURRAY, F.L.S., Keeper of Botany, British Museum. With 378 Illustrations. 8vo., 16s.

CROSS AND BEVAN.—CELLULOSE: an Outline of the Chemistry of the Structural Elements of Plants. With Reference to their Natural History and Industrial Uses. By CROSS and BEVAN (C. F. Cross, E. J. Bevan, and C. Beadle). With 14 Plates. Crown 8vo., 12s. net.

EDMONDS.—Works by HENRY EDMONDS, B.Sc., London.

   ELEMENTARY BOTANY, Theoretical and Practical. With 319 Diagrams and Woodcuts. Crown 8vo., 2s. 6d.

   BOTANY FOR BEGINNERS. With 85 Illustrations. Fcp. 8vo., 1s. 6d.

KITCHENER.—A YEAR'S BOTANY. Adapted to Home and School Use. With Illustrations by the Author. By FRANCES ANNA KITCHENER. Crown 8vo., 5s.

BOTANY—*Continued.*

LINDLEY AND MOORE.—THE TREASURY OF BOTANY; or, Popular Dictionary of the Vegetable Kingdom : with which is incorporated a Glossary of Botanical Terms. Edited by J. LINDLEY, M.D., F.R.S., and T. MOORE, F.L.S. With 20 Steel Plates and numerous Woodcuts. Two parts, fcp. 8vo., 12s.

McNAB.—CLASS-BOOK OF BOTANY. By W. R. McNAB.
MORPHOLOGY AND PHYSIOLOGY. With 42 Diagrams. Fcp. 8vo., 1s. 6d.
CLASSIFICATION OF PLANTS. With 118 Diagrams. Fcp. 8vo., 1s. 6d.

THOMÉ AND BENNETT.—STRUCTURAL AND PHYSIO-LOGICAL BOTANY. By Dr. OTTO WILHELM THOMÉ and by ALFRED W. BENNETT, M.A., B.Sc., F.L.S. With Coloured Map and 600 Woodcuts. Fcp. 8vo., 6s.

WATTS.—A SCHOOL FLORA. For the use of Elementary Botanical Classes. By W. MARSHALL WATTS, D.Sc. Lond. Crown 8vo., 2s. 6d.

AGRICULTURE AND GARDENING.

ADDYMAN.—AGRICULTURAL ANALYSIS. A Manual of Quantitative Analysis for Students of Agriculture. By FRANK T. ADDYMAN, B.Sc. (Lond.), F.I.C., Lecturer on Agricultural Chemistry, University College, Nottingham, etc. With 49 Illustrations. Crown 8vo., 5s. net.

COLEMAN AND ADDYMAN.—PRACTICAL AGRICULTURAL CHEMISTRY. For Elementary Students, adapted for use in Agricultural Classes and Colleges. By J. BERNARD COLEMAN, A.R.C.Sc., F.I.C., and FRANK T. ADDYMAN, B.Sc. (Lond.), F.I.C. With 24 Illustrations. Crown 8vo., 1s 6d. net.

LOUDON.—ENCYCLOPÆDIA OF AGRICULTURE; the Laying-out, Improvement, and Management of Landed Property; the Cultivation and Economy of the Productions of Agriculture. By J. C. LOUDON, F.L.S. With 1100 Woodcuts. 8vo., 21s.

SORAUER.—A POPULAR TREATISE ON THE PHYSIOLOGY OF PLANTS. For the use of Gardeners, or for Students of Horticulture and of Agriculture. By Dr. PAUL SORAUER. Translated by F. E. WEISS, B.Sc., F.L.S. With 33 Illustrations. 8vo., 9s. net.

WEBB.—Works by HENRY J. WEBB, Ph.D., B.Sc. (Lond.); late Principal of the Agricultural College, Aspatria.
ELEMENTARY AGRICULTURE. A Text-Book specially adapted to the requirements of the Science and Art Department, the Junior Examination of the Royal Agricultural Society, and other Elementary Examinations. With 34 Illustrations. Crown 8vo., 2s. 6d.
AGRICULTURE. A Manual for Advanced Science Students. With 100 Illustrations. Crown 8vo., 7s. 6d. net.

Scientific Works published by Longmans, Green, & Co. 27

WORKS BY JOHN TYNDALL, D.C.L., LL.D., F.R.S.

FRAGMENTS OF SCIENCE: a Series of Detached Essays, Addresses, and Reviews. 2 vols. Crown 8vo., 16s.

Vol. I.—The Constitution of Nature—Radiation—On Radiant Heat in Relation to the Colour and Chemical Constitution of Bodies—New Chemical Reactions produced by Light—On Dust and Disease—Voyage to Algeria to observe the Eclipse—Niagara—The Parallel Roads of Glen Roy—Alpine Sculpture—Recent Experiments on Fog-Signals—On the Study of Physics—On Crystalline and Slaty Cleavage—On Paramagnetic and Diamagnetic Forces—Physical Basis of Solar Chemistry—Elementary Magnetism—On Force—Contributions to Molecular Physics—Life and Letters of FARADAY—The Copley Medallist of 187c—The Copley Medallist of 1871—Death by Lightning—Science and the Spirits.

Vol. II.—Reflections on Prayer and Natural Law—Miracles and Special Providences—On Prayer as a Form of Physical Energy—Vitality—Matter and Force—Scientific Materialism—An Address to Students—Scientific Use of the Imagination—The Belfast Address—Apology for the Belfast Address—The Rev. JAMES MARTINEAU and the Belfast Address—Fermentation, and its Bearings on Surgery and Medicine—Spontaneous Generation—Science and Man—Professor VIRCHOW and Evolution—The Electric Light.

NEW FRAGMENTS. Crown 8vo., 10s. 6d.

CONTENTS.—The Sabbath—Goethe's 'Farbenlehre'—Atoms, Molecules, and Ether Waves —Count Rumford—Louis Pasteur, his Life and Labours—The Rainbow and its Congeners—Address delivered at the Birkbeck Institution on October 22, 1884—Thomas Young—Life in the Alps—About Common Water—Personal Recollections of Thomas Carlyle—On Unveiling the Statue of Thomas Carlyle—On the Origin, Propagation, and Prevention of Phthisis—Old Alpine Jottings—A Morning on Alp Lusgen.

LECTURES ON SOUND. With Frontispiece of Fog-Syren, and 203 other Woodcuts and Diagrams in the Text. Crown 8vo., 10s. 6d.

HEAT, A MODE OF MOTION. With 125 Woodcuts and Diagrams. Crown 8vo., 12s.

LECTURES ON LIGHT DELIVERED IN THE UNITED STATES IN 1872 AND 1873. With Portrait, Lithographic Plate, and 59 Diagrams. Crown 8vo., 5s.

ESSAYS ON THE FLOATING MATTER OF THE AIR IN RELATION TO PUTREFACTION AND INFECTION. With 24 Woodcuts. Crown 8vo., 7s. 6d.

RESEARCHES ON DIAMAGNETISM AND MAGNECRYSTALLIC ACTION; including the Question of Diamagnetic Polarity. Crown 8vo., 12s.

NOTES OF A COURSE OF NINE LECTURES ON LIGHT, delivered at the Royal Institution of Great Britain, 1869. Crown 8vo., 1s. 6d.

NOTES OF A COURSE OF SEVEN LECTURES ON ELECTRICAL PHENOMENA AND THEORIES, delivered at the Royal Institution of Great Britain, 1870. Crown 8vo., 1s. 6d.

LESSONS IN ELECTRICITY AT THE ROYAL INSTITUTION 1875-1876. With 58 Woodcuts and Diagrams. Crown 8vo., 2s. 6d.

THE GLACIERS OF THE ALPS: being a Narrative of Excursions and Ascents. An Account of the Origin and Phenomena of Glaciers, and an Exposition of the Physical Principles to which they are related. With numerous Illustrations. Crown 8vo., 6s. 6d. net.

FARADAY AS A DISCOVERER. Crown 8vo., 3s. 6d.

TEXT-BOOKS OF SCIENCE.

PHOTOGRAPHY. By Captain W. DE WIVELESLIE ABNEY, C.B., F.R.S., Director for Science in the Science and Art Department. With 155 Woodcuts. Price 3s. 6d.

THE STRENGTH OF MATERIALS AND STRUCTURES: the Strength of Materials as depending on their quality and as ascertained by Testing Apparatus; the Strength of Structures, as depending on their form and arrangement, and on the materials of which they are composed. By Sir J. ANDERSON, C.E., etc. With 66 Woodcuts. Price 3s. 6d.

RAILWAY APPLIANCES. A Description of Details of Railway Construction subsequent to the completion of Earthworks and Structures, including a short Notice of Railway Rolling Stock. By JOHN WOLFE BARRY, C.B., M.I.C.E. With 218 Woodcuts. Price 4s. 6d.

INTRODUCTION TO THE STUDY OF INORGANIC CHEMISTRY. By WILLIAM ALLEN MILLER, M.D., LL.D., F.R.S. With 72 Woodcuts. Price 3s. 6d.

QUANTITATIVE CHEMICAL ANALYSIS. By T. E. THORPE, F.R.S., Ph.D., Professor of Chemistry in the Royal College of Science, London. With 88 Woodcuts. Price 4s. 6d.

QUALITATIVE CHEMICAL ANALYSIS AND LABORATORY PRACTICE. By T. E. THORPE, Ph.D., D.Sc., F.R.S., Principal Chemist of the Government Laboratories, London, and M. M. PATTISON MUIR, M.A. With Plate of Spectra and 57 Woodcuts. Price 3s. 6d.

INTRODUCTION TO THE STUDY OF CHEMICAL PHILOSOPHY. The Principles of Theoretical and Systematic Chemistry. By WILLIAM A. TILDEN, D.Sc., London, F.R.S., Professor of Chemistry at the Royal College of Science. With 5 Woodcuts. With or without Answers to Problems. Price 4s. 6d.

ELEMENTS OF ASTRONOMY. By Sir R. S. BALL, LL.D., F.R.S., Lowndean Professor of Astronomy in the University of Cambridge. With 130 Woodcuts. Price 6s. 6d.

SYSTEMATIC MINERALOGY. By HILARY BAUERMAN, F.G.S., Associate of the Royal School of Mines. With 373 Woodcuts. Price 6s.

DESCRIPTIVE MINERALOGY. By HILARY BAUERMAN, F.G.S., etc. With 236 Woodcuts. Price 6s.

METALS, THEIR PROPERTIES AND TREATMENT. By C. L. BLOXAM and A. K. HUNTINGTON, Professors in King's College, London. With 130 Woodcuts. Price 5s.

PHYSICAL OPTICS. By R. T. GLAZEBROOK, M.A., F.R.S., Fellow and Lecturer of Trinity College, and Demonstrator of Physics at the Cavendish Laboratory, Cambridge. With 183 Woodcuts. Price 6s.

PRACTICAL PHYSICS. By R. T. GLAZEBROOK, M.A., F.R.S., and W. N. SHAW, M.A. With 134 Woodcuts. Price 7s. 6d.

Scientific Works published by Longmans, Green, & Co. 29

TEXT-BOOKS OF SCIENCE—*Continued.*

PRELIMINARY SURVEY. By Theodore Graham Gribble, Civil Engineer. Including Elementary Astronomy, Route Surveying, Tacheometry, Curveranging, Graphic Mensuration, Estimates, Hydrography, and Instruments. With 130 Illustrations, Quantity Diagrams, and a Manual of the Slide-Rule. Price 6s.

ALGEBRA AND TRIGONOMETRY. By William Nathaniel Griffin, B.D. Price 3s. 6d. Notes on, with Solutions of the more difficult Questions. Price 3s. 6d.

THE STEAM ENGINE. By George C. V. Holmes (Whitworth Scholar), Secretary of the Institution of Naval Architects. With 212 Woodcuts. Price 6s.

ELECTRICITY AND MAGNETISM. By Fleeming Jenkin, F.R.SS., L. & E., late Professor of Engineering in the University of Edinburgh. With 177 Woodcuts. Price 3s. 6d.

THE ART OF ELECTRO-METALLURGY, including all known Processes of Electro-Deposition. By G. Gore, LL.D., F.R.S. With 56 Woodcuts. Price 6s.

TELEGRAPHY. By W. H. Preece, C.B., F.R.S., V.P.Inst., C.E., Engineer-in-Chief, and Electrician, Post Office Telegraphs, and Sir J. Sivewright, M.A., K.C.M.G. With 258 Woodcuts. Price 6s.

THEORY OF HEAT. By J. Clerk Maxwell, M.A., LL.D., Edin., F.R.SS., L. & E. New Edition. With Corrections and Additions by Lord Rayleigh, Sec. R. S. With 38 Woodcuts. Price 4s. 6d.

TECHNICAL ARITHMETIC AND MENSURATION. By Charles W. Merrifield, F.R.S. Price 3s. 6d. Key, by the Rev. John Hunter, M.A. Price 3s. 6d.

THE STUDY OF ROCKS, an Elementary Text-Book of Petrology. By Frank Rutley, F.G.S., of Her Majesty's Geological Survey. With 6 Plates and 88 Woodcuts. Price 4s. 6d.

WORKSHOP APPLIANCES, including Descriptions of some of the Gauging and Measuring Instruments—Hand-Cutting Tools, Lathes, Drilling, Planing, and other Machine Tools used by Engineers. By C. P. B. Shelley, M.I.C.E. With an additional Chapter on Milling. By R. R. Lister. With 323 Woodcuts. Price 5s.

ELEMENTS OF MACHINE DESIGN. By W. Cawthorne Unwin, F.R.S., B.Sc., M.I.C.E. PART I. General Principles, Fastenings, and Transmissive Machinery. With 304 Woodcuts. Price 6s. PART II. Chiefly on Engine Details. With 174 Woodcuts. Price 4s. 6d.

STRUCTURAL AND PHYSIOLOGICAL BOTANY. By Dr. Otto Wilhelm Thomé, Rector of the High School, Cologne, and A. W. Bennett, M.A., B.Sc. F.L.S. With 600 Woodcuts and a coloured Map. Price 6s.

PLANE AND SOLID GEOMETRY. By H. W. Watson, M.A., formerly Fellow of Trinity College, Cambridge, Price 3s. 6d.

ADVANCED SCIENCE MANUALS.

⁎ *Written specially to meet the requirements of the ADVANCED STAGE of Science Subjects as laid down in the Syllabus of the Directory of the SCIENCE AND ART DEPARTMENT, SOUTH KENSINGTON.*

BUILDING CONSTRUCTION. By the Author of 'Rivington's Notes on Building Construction'. With 385 Illustrations and an Appendix of Examination Questions. Crown 8vo., 4s. 6d.

THEORETICAL MECHANICS. Solids, including Kinematics, Statics, and Kinetics. By A. THORNTON, M.A., F.R.A.S., With 220 Illustrations, 130 Worked Examples, and over 900 Examples from Examination Papers, etc. Crown 8vo., 4s. 6d.

HEAT. By MARK R. WRIGHT, Hon. Inter. B.Sc. Lond. With 136 Illustrations and numerous Examples and Examination Papers. Crown 8vo., 4s. 6d.

LIGHT. By W. J. A. EMTAGE, M.A. With 232 Illustrations. Cr. 8vo., 6s.

MAGNETISM AND ELECTRICITY. By ARTHUR WILLIAM POYSER, M.A. With 317 Illustrations. Crown 8vo., 4s. 6d.

INORGANIC CHEMISTRY, THEORETICAL AND PRACTICAL. A Manual for Students in Advanced Classes of the Science and art Department. By WILLIAM JAGO, F.C.S., F.I.C. With Plate of Spectra and 78 Woodcuts. Crown 8vo., 4s. 6d.

GEOLOGY: a Manual for Students in Advanced Classes and for General Readers. By CHARLES BIRD, B.A. (Lond.), F.G.S. With over 300 Illustrations, a Geological Map of the British Isles (coloured), and a set of Questions for Examination. Crown 8vo., 7s. 6d.

HUMAN PHYSIOLOGY: a Manual for Students in advanced Classes of the Science and Art Department. By JOHN THORNTON, M.A. With 268 Illustrations, some of which are Coloured, and a set of Questions for Examination. Crown 8vo., 6s.

PHYSIOGRAPHY. By JOHN THORNTON, M.A. With 6 Maps, 180 Illustrations, and Coloured Plate of Spectra. Crown 8vo., 4s. 6d.

AGRICULTURE. By Dr. HENRY J. WEBB, Ph.D., B.Sc. With 100 Illustrations. Crown 8vo., 7s. 6d. net.

ELEMENTARY SCIENCE MANUALS.

PRACTICAL, PLANE, AND SOLID GEOMETRY, including Graphic Arithmetic. By I. H. MORRIS. Fully Illustrated with Drawings prepared specially for the book. Crown 8vo., 2s. 6d.

GEOMETRICAL DRAWING FOR ART STUDENTS. Embracing Plane Geometry and its Applications, the Use of Scales, and the Plans and Elevations of Solids, as required in Section I. of Science Subject I. By I. H. MORRIS. Crown 8vo., 1s. 6d.

TEXT-BOOK ON PRACTICAL, SOLID, OR DESCRIPTIVE GEOMETRY. By DAVID ALLAN LOW (Whitworth Scholar). Part I. Crown 8vo., 2s. Part II. Crown 8vo., 3s.

AN INTRODUCTION TO MACHINE DRAWING AND DESIGN. By DAVID ALLAN LOW (Whitworth Scholar). With 97 Illustrations and Diagrams. Crown 8vo., 2s.

Scientific Works published by Longmans, Green, & Co. 31

ELEMENTARY SCIENCE MANUALS—*Continued.*

BUILDING CONSTRUCTION. By EDWARD J. BURRELL, Second Master of the Technical School of the People's Palace, Mile End. With 308 Illustrations and Working Drawings. Crown 8vo., 2s. 6d.

AN ELEMENTARY COURSE OF MATHEMATICS. Containing Arithmetic; Euclid (Book I., with Deductions and Exercises); and Algebra. Crown 8vo., 2s. 6d.

THEORETICAL MECHANICS. Including Hydrostatics and Pneumatics. By J. E. TAYLOR, M.A., B.Sc. With numerous Examples and Answers, and 175 Diagrams. Crown 8vo., 2s. 6d.

THEORETICAL MECHANICS—SOLIDS. By J. E. TAYLOR, M.A., B.Sc. With 163 Illustrations, 120 Worked Examples, and over 500 Examples from Examination Papers, etc. Crown 8vo., 2s. 6d.

THEORETICAL MECHANICS—FLUIDS. By J. E. TAYLOR, M.A., B.Sc. With 122 Illustrations, numerous Worked Examples, and about 500 Examples from Examination Papers, etc. Crown 8vo., 2s. 6d.

A MANUAL OF MECHANICS: an Elementary Text-Book for Students of Applied Mechanics. With 138 Illustrations and Diagrams, and 188 Examples taken from the Science Department Examination Papers, with Answers. By T. M. GOODEVE, M.A. Fcp. 8vo., 2s.

SOUND, LIGHT, AND HEAT. By MARK R. WRIGHT. Hon. Inter. B.Sc., London. With Examples, Examination Papers, and 165 Illustrations. Crown 8vo., 2s. 6d.

PHYSICS. Alternative Course. By MARK R. WRIGHT, Hon. Inter. B.Sc., London. With Examples, Examination Papers, and 242 Illustrations. Crown 8vo., 2s. 6d.

ELEMENTARY PRACTICAL CHEMISTRY: a Laboratory Manual for Use in Organised Science Schools. By G. S. NEWTH, F.I.C., F.C.S. Demonstrator in the Royal College of Science, London; Assistant Examiner in Chemistry, Science and Art Department. With 108 Illustrations and 254 Experiments. Crown 8vo. Price 2s. 6d.

ELEMENTARY PRACTICAL PHYSICS: a Laboratory Manual for Use in Organised Science Schools. By W. WATSON, B.Sc. Demonstrator in Physics in the Royal College of Science, London; Assistant Examiner in Physics, Science and Art Department. With 119 Illustrations and 193 Exercises. Crown 8vo. Price 2s. 6d.

MAGNETISM AND ELECTRICITY. By A. W. POYSER, M.A. With Examination Papers and 235 Illustrations. Crown 8vo., 2s. 6d.

PROBLEMS AND SOLUTIONS IN ELEMENTARY ELECTRICITY AND MAGNETISM. Embracing a Complete Set of Answers to the South Kensington Papers for the Years 1885-1894, and a Series of Original Questions. By W. SLINGO and A. BROOKER. With 67 Illustrations. Crown 8vo., 2s.

INORGANIC CHEMISTRY, THEORETICAL AND PRACTICAL. With an Introduction to the Principles of Chemical Analysis. By WILLIAM JAGO, F.C.S., F.I.C. With 63 Woodcuts and numerous Questions and Exercises. Fcp. 8vo., 2s. 6d.

AN INTRODUCTION TO PRACTICAL INORGANIC CHEMISTRY. By WILLIAM JAGO, F.C.S., F.I.C. With Illustrations. Crown 8vo., 1s. 6d.

PRACTICAL CHEMISTRY: the Principles of Qualitative Analysis. By WILLIAM A. TILDEN, D.Sc. With Illustrations. Fcp. 8vo., 1s. 6d.

ELEMENTARY CHEMISTRY, Inorganic and Organic. By WILLIAM S. FURNEAUX. With Examination Questions, and 65 Illustrations. Crown 8vo., 2s. 6d.

ORGANIC CHEMISTRY: the Fatty Compounds. By R. LLOYD WHITELEY, F.I.C., F.C.S. With 45 Illustrations. Crown 8vo., 3s. 6d.

ELEMENTARY GEOLOGY. By CHARLES BIRD, B.A., F.G.S. With Coloured Geological Map of the British Islands, and 247 Illustrations. Crown 8vo., 2s. 6d.

32 *Scientific Works published by Longmans, Green, & Co.*

ELEMENTARY SCIENCE MANUALS—*Continued*.

HUMAN PHYSIOLOGY. By WILLIAM S. FURNEAUX. With 218 Illustrations. Crown 8vo., 2s. 6d.

BIOLOGY. By JOHN BIDGOOD, B.Sc. With 226 Illustrations. Crown 8vo., 4s. 6d.

ELEMENTARY BOTANY, THEORETICAL AND PRACTICAL. By HENRY EDMONDS, B.Sc., London. With 319 Woodcuts. Crown 8vo., 2s. 6d.

METALLURGY. By E. L. RHEAD, Lecturer on Metallurgy at the Municipal Technical School, Manchester. With 94 Illustrations. Crown 8vo., 3s. 6d.

STEAM. By WILLIAM RIPPER, Member of the Institution of Mechanical Engineers, Professor of Mechanical Engineering in the Sheffield Technical School. With 142 Illustrations. Crown 8vo., 2s. 6d.

ELEMENTARY PHYSIOGRAPHY. By JOHN THORNTON, M.A. With 12 Maps and 247 Illustrations. Crown 8vo., 2s. 6d.

AGRICULTURE. By HENRY J. WEBB, Ph.D., B Sc. (Lond.). Late Principal of the Agricultural College, Aspatria. With 34 Illustrations. Crown 8vo., 2s. 6d.

HYGIENE. By J. L. NOTTER, M.A., M.D., Fellow and Member of Council of the Sanitary Institute of Great Britain, Examiner in Hygiene, Science and Art Department; Examiner in Public Health in the University of Cambridge and in the Victoria University, Manchester; and R. H. FIRTH, F.R.C.S, Assistant Professor of Hygiene, in the Army Medical School, Netley. With 95 Illustrations. Crown 8vo., 3s. 6d.

THE LONDON SCIENCE CLASS-BOOKS.

Edited by G. CAREY FOSTER, F.R.S., and by Sir PHILIP MAGNUS, B.Sc., B.A., of the City and Guilds of London Institute.

ASTRONOMY. By Sir ROBERT STAWELL BALL, LL.D., F.R.S. 41 Diagrams. 1s. 6d.

MECHANICS. By SIR ROBERT STAWELL BALL, LL.D., F.R.S. 89 Diagrams. 1s. 6d.

THE LAWS OF HEALTH. By W. H. CORFIELD, M.A., M.D., F.R.C.P. With 22 Illustrations. 1s. 6d.

MOLECULAR PHYSICS AND SOUND. By FRED. GUTHRIE, F.R.S. 91 Diagrams. 1s. 6d.

GEOMETRY, CONGRUENT FIGURES. By O. HENRICI, Ph.D., F.R.S. With 141 Diagrams. 1s. 6d.

ZOOLOGY OF THE INVERTEBRATE ANIMALS. By ALEXANDER MACALISTER, M.D. With 59 Diagrams. 1s. 6d.

ZOOLOGY OF THE VERTEBRATE ANIMALS. By ALEXANDER MACALISTER, M.D. With 77 Diagrams. 1s. 6d.

HYDROSTATICS AND PNEUMATICS. By Sir PHILIP MAGNUS, B.Sc., B.A. 79 Diagrams. 1s. 6d. (To be had also *with Answers*, 2s.). The Worked Solutions of the Problems. 2s.

BOTANY. Outlines of the Classification of Plants. By W. R. McNAB, M.D. With 118 Diagrams. 1s. 6d.

BOTANY. Outlines of Morphology and Physiology. By W. R. McNAB, M.D. With 42 Diagrams. 1s. 6d.

THERMODYNAMICS. By RICHARD WORMELL, M.A., D.Sc. 41 Diagrams.

5,000/11/96.

www.ingramcontent.com/pod-product-compliance
Lightning Source LLC
Chambersburg PA
CBHW020136170426
43199CB00010B/762